God's Salesman

The world traveler and elder statesman at 76: active, vital, and still in demand. A warm smile and lack of affectation endeared Peale to his audiences.

God's Salesman

NORMAN VINCENT PEALE & THE POWER OF POSITIVE THINKING

Carol V. R. George

OXFORD UNIVERSITY PRESS
New York Oxford

Oxford University Press

Oxford New York Toronto
Delhi Bombay Calcutta Madras Karachi
Kuala Lumpur Singapore Hong Kong Tokyo
Nairobi Dar es Salaam Cape Town
Melbourne Auckland Madrid

and associated companies in
Berlin Ibadan

First published in 1993 by Oxford University Press, Inc.,
200 Madison Avenue, New York, New York 10016

First issued as an Oxford University Press paperback, 1994

Library of Congress Cataloging-in-Publication Data
George, Carol V. R.
God's Salesman : Norman Vincent Peale / Carol V. R. George
p. cm. (Religion in America series) includes bibliographical references and index.
ISBN 0-19-507463-7
ISBN 0-19-508915-4 (pbk.)
1. Peale, Norman Vincent. 1898– . 2. Reformed Church in America—Clergy—Biography. 3. Reformed Church—United States—Clergy—Biography. 4. Peace of mind—Religious aspects—Christianity—History of doctrines—20th century. I. Title. II. Series: Religion in America series (Oxford University Press) 285.7′092—dc20 [B] 92-16176

2 4 6 8 10 9 7 5 3 1

Printed in the United States of America

Especially for David and Rosalind
and for
Bill and Robin with love

Preface

At the noisy commercial intersection of Twenty-ninth Street and Fifth Avenue in Manhattan, on the northwest corner, is the somber grey stone Marble Collegiate Church, for fifty-two years the ecclesiastical home of the Reverend Dr. Norman Vincent Peale. Next to the Fifth Avenue entrance to the church and embedded in its stone is a plaque of recent origin, which reads:

> Dr. Norman Vincent Peale, D.D., L.L.D.
> Minister 1932–1984
>
> From the pulpit of this venerable church, this man with matchless eloquence and person to person persuasiveness sent forth a message that has circled the globe: he taught that positive thinking when applied to the power of the Christian message could not only overcome all difficulties, but also bring about triumphant lives. Let all who worship here, then, be open to Dr. Peale's momentous message.

This study examines the development and impact of Peale's global message, which the plaque correctly identifies with the extraordinary talents of the man himself. The general theme of the work is the interaction of religion and culture, here considered through the version of "Popular Christianity" which emerged from the relationship of Peale, his audience, and his message. Popular religion, a subject of considerable recent interest to historians of religion, is in fact the frame that defines and holds in place the cultural efflorescence I identify as Pealeism.[1]

Although the essential theme of the work is the evolution of a movement, its subject is appropriately Peale. In keeping with the nature of popular religious movements, Peale was not only its compelling, sometimes charismatic, leader but was himself perceived as the incarnation of its most basic values. Much like other engaging popular figures in our own day, such as Ronald

1. Sydney Ahlstrom's *A Religious History of the American People* (New Haven, CT: Yale University Press, 1972) was the ground-breaking work that identified Harmonial Religion and set it in context. The studies of popular religion that I found most helpful were Peter Williams' *Popular Religion in America* (Chicago: University of Illinois Press, 1989); Nathan Hatch's *The Democratization of American Christianity* (New Haven, CT: Yale University Press, 1989); and Jon Butler's *Awash in a Sea of Faith* (Cambridge, MA: Harvard University Press, 1990). As Williams' discussion makes abundantly clear, students of popular religion borrow heavily from the work of anthropologists, particularly ethnographers, as I have done.

Reagan on the right and Jesse Jackson on the left, Peale captured the language, symbol system, and values of his followers and shaped them into a unique message.

This, then, is the story of one modern version of a familiar historic tale, namely, the reappearance in mainstream culture of the intellectually unsophisticated religious beliefs of the people, or vernacular religion, as advanced by a self-defined grass-roots leader. It is a uniquely American tale, for Peale's credo was premised on myths deeply embedded in the national culture. It took for granted the reality of a chosen people in a chosen nation, unquestionably marked for social mobility and enduring national greatness. The major events of this modern saga, played out in the years surrounding World War II, and especially gripping public attention during the Cold War period of the 1950s, made Peale's name a household word. The high water mark for the phenomenon of Pealeism can be located in the two decades following World War II—and thus part of what historian William McLoughlin termed "The Fourth Great Awakening" of religion—while still leaving a mark on American life that remains very visible today. Characteristically for popular religion, it is a story of struggle, the bitter struggle between a populist insurgency and the establishment it challenged, which in this case assumed particularly harsh overtones because of Peale's professional identification with that establishment, and consequently his presumed apostasy.

It is, moreover, a story about the quotidian, the daily concerns and stresses of his bourgeois constituency—largely traditional Protestants of an evangelical bent—which felt its personal and spiritual needs neglected by its mainstream mentors as they championed other, more socially sensitive causes. His followers were looking for answers, frequently for ways out of particular personal crises, and Peale provided uncomplicated responses that relied very little on the professional expertise of the establishment and fundamentally on individual initiative. He reminded them that the answer to their troubles lay within themselves, in the divine energy stored within the unconscious, which they had only to tap through affirmative prayer and positive thinking. His was a message of personal empowerment to an audience that was predominantly female, Protestant, and lower middle class: The female majority in the movement, usually constituting the faceless multitude, was generally overshadowed by the highly visible minority of businessmen who sustained the program financially and came to represent the entire organism in the public debate. Curiously, the minister known affectionately as "God's Salesman," served a national congregation made up largely of women.

Popular religion has typically been regarded by intellectuals as suspect, superstitious, and antinomian, a danger to the established order, forgetting that most modern religious organizations had their inceptions not in academies or seminaries but in fields and upper rooms. One potential evolutionary route for popular religion, therefore, is through respectability into the mainstream. Another, and much more likely, path is one that eventually fades from the cultural scene following the death of the leader and sometimes his or her successor, but almost always after leaving evidence of its agenda on the host society. Peale's version of Popular Christianity shares many of the characteristics associated with religious populism generally. With popular religion, the basic imagery tends to derive from ancient myths, primarily those that emphasize the concept of oneness and primal bonding and that employ traditional language and symbolic forms to make these themes relevant to the issues of the present age.[2] The emphasis on personal bonding reduces the importance of the institutionalization of religion, including rituals, priestcraft, and systematized belief. Popular religion has usually emerged from rural soil, although in Peale's case the roots finally took hold in an urban environment.[3]

Consistent with a view of oneness and bonding, popular religion subscribes to a holistic view of creation, one that links mind, body, and spirit in a cosmic whole. Its message is pragmatic, self-consciously, even dogmatically, empirical, syncretic, and nonsystematic, expressed in the language of the masses and communicated through a technological vanguardism, often experimental and relatively sophisticated, designed to reach large numbers. Given the nature of its origins, popular religion tends to be politically conservative though personally optimistic, suspicious of bureaucracy and more comfortable with decentralized, even autonomous, groups. As with the rest of its worldview, Christian forms of popular religion tend to express a Christology that is nonsystematic, alternately emphasizing a view of Jesus as personal friend and companion and a Christ conceived as synonymous with God or Divine Mind.

Just what to call the variant of popular religion with which Peale's movement is identified has been a problem. Peale himself called it practical Christianity, claiming that he had simply taken the "old fashioned gospel story" and applied it through a modern idiom to the practical problems of daily living. But it is the old-fashioned gospel with a significant adaptation: In its emphasis on a personally manipulable unconscious force, which Peale called "thought power" or positive thinking, it conceived of a link between the individual and divine energy. That theme of cosmic oneness, evident in ancient myths of bonding and wholeness, enjoyed a rebirth in America in the nineteenth century through the metaphysical contributions of the Transcendentalists and the Sage of Concord, Ralph Waldo Emerson. Peale's movement is kin to this American product, as it evolved through the Transcendentalist-New Thought-Christian Science tradition on the one hand

2. "Popular Religion," in *Encyclopedia of Religion*, Mircea Eliade, Editor-in-Chief (New York: Macmillan, 1987), Vol. 11. See also Catherine Albanese, *Nature Religion in America* (Chicago: University of Chicago Press, 1990).

3. This emphasis is particularly vivid in Peale's 1991 book, *This Incredible Century* (Wheaton, IL: Tyndale House), which is his reminiscence of a life that began in 1898.

and the mesmerist-mind cure-psychoanalytic experience on the other. This continuing sectarian tradition, which honors the wholeness of nature, has much in common with what Catherine Albanese has called "Nature Religion,"[4] and others have perceived as a kind of grass-roots folk religion. Clearly identified by Peale with Conservative politics, the movement and its founder became the focus of an ongoing debate with liberals, who frequently aligned it with an experience they indiscriminately associated with "pathological yahoos."[5]

Peale himself has disliked the term "Pealeism" for his movement, rightly suspecting that it is meant as a way of characterizing a development outside the framework of traditional Protestantism and containing hints of the cultish aspect of personality-based belief groups. For despite his awareness of the unique quality of his ministry, Peale has consistently identified it with the "old fashioned gospel" of mainstream evangelicalism. As he saw it, it was not he who had strayed from the traditional beliefs but the liberals, while he was simply clothing the old-time concepts in a newer style.

This book examines the ingredients of Peale's movement and of its interaction with mainstream American religion, essentially liberal Protestantism. I have also attempted to identify those developments in the movement—such as its obvious politicization, its rapport with secular trends, its appeal for middle-aged, middle-class women, and its rapprochement with revitalized evangelicalism—that enabled it during Peale's ministry finally to penetrate the mainstream.

Geneva, New York C. G.
June 1992

4. Albanese, *Nature Religion in America*.

5. The colorful phrase is taken from a review essay by Michael Kazin, "The Grass-Roots Right: New Histories of U.S. Conservatism in the Twentieth Century," in *The American Historical Review*, February 1992, 97:136–55.

Acknowledgments

Among the millions of people who have supported Norman Vincent Peale's message of positive thinking and what he terms practical Christianity, salesmen have held a special place. Over the years Peale has addressed hundreds of meetings of sales organizations, the awards and plaques lining the walls leading to his office the visible manifestation of the mutual regard he and they had for each other. The real significance of these honors, however, attaches to the image Peale had of salesmen as quintessential Americans, whose very livelihood depended on an optimistic hold on the future, entrepreneurs of the self who had to believe in their own ability to make the next sale. For Peale, these people—in his day necessarily men—were the embodiment of what his message was about: They seemingly premised their lives on a positive attitude, a belief in themselves and their product, and an awareness that continued success in their careers required the constant revitalization of the internal forces that drove their customer appeals. Reflections of the title of Peale's 1938 book, *You Can Win*, they offered a small but telling example of his simple message—namely, that individuals in all walks of life possessed abundant inner resources to confront the daily challenges that awaited them.

His expression of practical Christianity, banal and meaningless to the sophisticated, has nevertheless held a prominent place in popular culture and the American mainstream at least since 1952. That was the year that Peale's best-selling *The Power of Positive Thinking* made him an international celebrity and his message the resource for millions of Americans searching for solutions to personal problems and status anxieties. It also made him a target in the historic struggle between intellectuals and popular culture.

This book examines Peale's message as well as the man, and attempts to evaluate both within the context of American culture in the decades following World War II. My hope has been to try to make sense of issues that have been important to me as I observed as well as participated in mainstream institutional religious life. People of my generation, alternately troubled by signs of the decline of the liberal witness and cheered by hopeful new gleams of light, have tended to harbor what might most generously be termed an ambivalent attitude toward groups that posed a challenge on the cultural margins, as Peale's constituency once did. Now marginal ourselves, we have also become more willing to take seriously the religious expressions and cultural implications of the kind of popular religion we once dismissed as cheap grace and spiritual fluff. An important element in Peale's challenge to the religious establishment was just that sort of demand for recognition. In some quarters, at least, it has been successful, as the new revisionist scholarship on populist conservatism can attest.

The study has been over a decade developing. Initiated in December 1980 after Dr. Peale granted me what was essentially a courtesy interview about another subject, it has survived major alterations in my personal life while my professional world has remained fortunately intact. Although in no sense is this an authorized biography, the book could not have been undertaken without Dr. Peale's cooperation. Not only did he grant me sole access to what was then an 850-box private collection, the Norman Vincent Peale Manuscript Collection at Syracuse University, but he was generous with the time he made available for interviews. He made it possible for me to meet with other key people in his organization, and never inhibited the free flow of information. Without his help and that of his wife, Ruth Stafford Peale, I would not have been able to undertake this work.

As a result, I have felt especially connected to them as I developed the book. While all historians feel a responsibility to recapture the past as faithfully as possible, those who work with a living subject experience a compelling, sometimes awesome, sense of the weight of objective history. I have tried to tell the story from "inside" the movement, utilizing an anthropological model associated with Clifford Geertz, an approach that promotes objectivity while not eliminating the need for moral judgment and evaluation. Despite my goal of objectivity, however, I suspect that Dr. Peale, an especially sensitive man, will be unsettled with some of what I have written. This potential was never far from my consciousness as I wrote the book, yet when faced with the choice between telling the story as the evidence suggested and having his feelings guide my judgment, I chose the former. My hope has been that Dr. Peale appreciates the need for intellectual honesty and accepts it, however reluctantly, as the price one pays for holding a place in history.

Other members of his organization have facilitated my work. Most of my contacts with the Peales have occurred in or near his Pawling, New York, headquarters, formerly the Foundation for Christian Living (FCL) and since October 1991 the Peale Center for Christian Living. I have also met with the Peales at their apartment in New York City and in 1981 joined them in an FCL-sponsored conference in Madrid. All of these arrangements were made possible by Dr. Peale's personal secretaries at FCL, first Evelyn Yegella and then Sybil Light. Typical of his FCL staff, they were eager to be helpful, predictably efficient, and always pleasant. Because I also used a clipping collection housed at the FCL center, I spent a considerable amount of time there and consequently came to know many of the staff on a casual, informal basis. It is difficult to single out individuals for special thanks, although Kenneth Winslow, Ann Munro, Rocco Murano, Eric Fellman, and Mark Lambert are individuals I especially remember for their willingness to be helpful.

At *Guideposts* magazine, the literary link for Peale's global ministry, I profited from conversations with John Beech and Van Varner and from their efforts to see that I had a complete run of *Guideposts'* publications. In New York, Dr. Peale's successor at Marble Collegiate Church, Dr. Arthur Caliandro, talked with me on several occasions and even responded to questions on the phone as he was about to depart for his summer vacation. I had a particularly warm reception at the Institutes for Religion and Health, where I had an open and lengthy conversation with the then-director of training, Stephen Pritchard, and attended the graduation exercises for its professional interns at the Blanton-Peale Institute. On that occasion, I enjoyed the company of Arthur Gordon, the bright, atypically irreverent Peale biographer and former *Guideposts* editor. The Peales and their personal staff made all these meetings possible, and I am indebted to them for their help.

I also owe a large debt of thanks to the archival staff at Syracuse University Library, where the enormous Peale manuscript collection is housed. Carolyn A. Davis, Director, and her staff made my long stay in the manuscripts room pleasant and productive.

At my own institution, Hobart and William Smith Colleges, the reference staff, particularly Gary Thompson and then Michael Hunter, diligently tracked down even the remotest shred of data I requested. I have been constantly amazed by both their ingenuity and their speed in getting material back to me. At the Colleges, too, my research activities and leaves from campus posed additional burdens for my colleagues, and I am eternally indebted to them for their patience, help, and good humor. During the long period of developing the work, provosts have come and gone, but they have been uniformly supportive of my efforts: Harmon Dunathon, Minor Myers, and Sheila Bennett. Former president of the Colleges Carroll Brewster assisted with my research on his grandfather, the Reverend Elwood Worcester, by making an important clipping collection available, introducing me to his aunt, who was Elwood's daughter, and sharing personal memories with me. My special thanks go to my history department colleagues, in particular to my American history friends Bob Huff, Dan Singal, and Jim Crouthamel. They have taken up the slack for me many times, and I am grateful to them.

For those who write and teach, finding ways to fund time for writing is always a challenge. I have been fortunate to be on the receiving side of grants at crucial times in the development of the work. Hobart and William Smith Colleges have been a continuing source of support in terms of summer awards for research and in helping to make possible the acceptance of external awards. I was also encouraged to receive a summer stipend from the National Endowment for the Humanities at an early stage in the project. Later, at the writing stage, I was pleased to receive from NEH a Fellowship for Independent Scholarship and Research.

Early direction for the project came as a result of an NEH summer seminar at Harvard Divinity School directed by William Hutchison. Professor Hutchison gently nudged my focus while also suggesting important resources. He has been an invaluable guide, friend, and reference ever since.

Oxford University Press editor Cynthia Read has been all one could ask for in an editor. She has been enthusiastic about the manuscript from the first, while also being a careful and knowledgeable critic.

Over the years, friends and family have borne with endless Peale conversations and have shared with me Peale information, as well as love and support. My longtime friend Jane Donegan unknowingly shared a typical Peale boost when she said over lunch as I initially proposed the topic, "Of course, you'll do it, and do it well." Coming from a friend whose candor I value, it was the kind of motivation I needed to get going on a long and complicated project, as she probably well understood at the time. Betty Bone Schiess, Lois Black, and Judy Hill Fleming were attentive listeners and important sources of support during rough times. I discovered an able bibliographer and library sleuth in Amie Seymour. And Rachel Harper, smart, skilled, and persistent, finally arranged all the bibliographical material and rechecked footnotes. My good friend Richard Newman, from his vantage point at the New York Public Library, kept me current with developing Peale news and read and commented on the manuscript. He has also been an important source of uncommon good sense.

As is always the case, however, the greatest burden was imposed on my family, and to them it is difficult to express fully the extent of my gratitude. In the early years, my most enthusiastic supporter was my husband Bill, who, it seemed to me, always overestimated my talents. He thought it was a story that needed to be told, and that I was able to tell it. His death in 1988 left me without his companionship, support, and advice. It also meant that in a now single-parent family, David and Rosalind, teenagers still at home, found their lives significantly restructured with much more independence expected of them. They have been magnificent. In addition to coping with their own special adolescent-related anxieties, they have helped me manage a complicated household, took for granted my constant place in front of the computer, and added zest and spirit to my life. My son Bill and his wife Robin have helped in other, more grown-up ways, offering amateur psychotherapy when I got discouraged and helping to manage the details of my house, garden, dogs, and budget. I love them for being who they are, and for tolerating my impositions on their affection. Their confidence has kept me going.

Contents

God's Salesman

Mind is the only reality, of which men and all other natures are better or worse reflectors.

Ralph Waldo Emerson
"The Transcendentalist"
1841

I know I can, I think I can, I can, I can, I can.

Watty Piper
The Little Engine That Could
1990

Introduction

To many thoughtful Americans in the closing years of the twentieth century, the statistical evidence on the health of the culture speaks of a giant, fatal cancer, steadily and inexorably destroying the quality of life that was familiar and comfortable to anyone born before the Korean War. The data, regularly published in the press as if to titillate morbid sensibilities, confirm that one's personal experience with social disruption is general: Marriages and families seem increasingly fragile; children of all ages appear more at risk; the elderly live longer, hollower lives; ethnic groups battle each other for an even smaller part of the national pie; women and men weary of ever understanding each other; and national resources and prestige decline as the business community grows paralyzed from competition, complacency, and cultural pollution. There are people in middle life to whom it seems difficult to remember a brighter day, when life promised hope, a future of meaningful connections, and children had a right to large dreams. Who could think positively about the future?

Yet there was, and is, a mirror image to this scenario, one peopled by actors deeply engaged in efforts to improve their lives. Among these other Americans thinking positively about the future is far from a joke; the future, at least one's own part in it, can deliver millennial expectations. Some individuals are old hands at practicing the techniques of positive thinking, while others are new, eager students, their information coming from popular books, therapists, friends, or positive thinking seminars. They fit no precise demographic profile, although they are obviously not the poor and less privileged of the nation. They are senior citizens, athletes and public performers, New Age believers and youthful evangelicals, substance abusers, the terminally ill. And they seem heedless of the social critics who scold that positive thinking is a form of narcissism, an example of mental manipulation that is personally useless and culturally destructive because it is radically autonomous. The response of the converted has been "it works."

When a 1984 Gallup poll revealed that the same percentage of people were involved with positive thinking seminars as held membership in the Methodist Church it did not take the leaders of America's mainstream religious bodies by surprise.[1] Help seekers for some time had been trying seemingly new alternatives, testing nontraditional sources of support in their efforts to find answers to life's crises. The evidence showed that over the preceding two decades major realignments in religious affiliation had occurred, simultaneous with the proliferation of hundreds of new religious bodies. The soul hungry confronted a smorgasbord of new and unusual religious choices. Although positive thinking—metaphysical mental science descended through Christian Science and New Thought—had acquired a kind of parity with Methodism, it was not one of the "new" post-World War II religious creations. Its post-war popularizer, the Reverend Dr. Norman Vincent Peale, had enabled it finally to gain a place in the religious mainstream, but it had been a definable force in the spiritual panoply since before the turn of the century.

Social commentators, both inside and outside religious circles, who had been tracking these developments for several decades, were troubled by the postwar surge in unconventional forms of piety and the proliferation of sectarian bodies. Some feared they portended a dark age of irrationalism and religious deviance. At the very least they suggested that the reasonable religious liberalism of the historic denominations was under siege, replaced, perhaps, by narcissistic or emotional attachment to either popular occultic forms or an unyielding fundamentalist orthodoxy. Was it possible that positive thinking, for example, had anything in common with conventional religious belief? Was the news from the religious front to be received as a hopeful harbinger of a global renaissance of spiritual renewal, as in the Third World? Or was it the unfortunate but predictable outcome of modernization, a note on the "Europeanization" of traditional religion in America?

These larger cultural questions are necessarily hinged to an examination of the ministry and career of Norman Vincent Peale (1898–). Born in the bosom of religious orthodoxy, Peale became the controversial postwar symbol of the nation's shift in spiritual priorities. Preacher, author, editor, public personality, entrepreneur, and religious innovator, he was a religious populist whose challenge to the ecclesiastical establishment was especially troublesome to its defenders because of his own mainstream credentials and reputation. Educated at a Methodist college and seminary, he was the minister of the historic Collegiate Church of the Reformed Church in America, on Fifth Avenue in New York City. Was he a heretic to the sacred traditions he was supposed to uphold? Or was he a prophet exposing the neglect and pastoral disregard of priests who enjoyed only the challenge of political activism and social change? Was his message a magnet, rescuing for the church the uncommitted, the marginal, the agnostic—as Peale hoped—or was it—as its critics alleged—a secularized distortion of ancient Christian beliefs, a form of cheap grace, a shallow faith marketed to the masses? The controversy that swirled around these questions became a defining quality of Peale's ministry.

The debate over populist expressions of culture, what might be termed the "democratization" of culture, was hardly limited to the heyday of Peale's ministry. In various forms, it has gained national attention from the nineteenth century into our own day: The popularity of Allan Bloom's book, *The*

Closing of the American Mind, and the academic furor to which it contributed over the nature of the canon, reveal the continuing struggle in American life to isolate a distinctive, homogeneous "high" culture from a pluralistic, broad-based "popular" or lowbrow culture. In his penetrating analysis of cultural hierarchy in America, Lawrence Levine has noted:

> There is, finally, the same sense that culture is something created by the few for the few, threatened by the many, and imperiled by democracy; the conviction that culture cannot come from the young, the inexperienced, the untutored, the marginal; the belief that culture is finite and fixed, defined and measured, complex and difficult of access, recognizable only by those trained to recognize it, comprehensible only to those qualified to comprehend it.[2]

In a sense, the conflict offers new evidence for the observation made by Richard Hofstadter that there is a tendency among academics to endorse both high culture and democratization, but then to complain bitterly when democratization chips away at high culture and barbarizes it.[3]

Some of the strongest social criticism is reserved for just those subjectively oriented therapeutic and psychological practices most in evidence in Peale's work and most likely to be found among the middle class. Christopher Lasch, perhaps the best-known and most insistent of these critics, has hammered away at the "helping professionals" and their associated therapeutic prescriptions, contending that the most vital issue today is not self-fulfillment but survival.[4] The conservative social analyst Philip Rieff has claimed that the entire modern ethos might be described as a kind of "Therapeutic Culture," with the representative "Psychological Man" engaged in a fruitless search to find in therapy the reassurances and moral insight once sought in religion.[5]

The pessimism that emerges from these jeremiads is countered by the sort of perspective that emerges from works such as those by Levine, which argue that culture is dynamic and that at this point in time is no nearer an apocalyptic end than at any other. Proponents of this view see two related forces at work. One is the process of modernization itself, which in its shifting emphasis from the traditional community to the individual inevitably creates a focus on the plastic, psychological self.[6] The other, which reiterates the emphasis on cultural inevitability, sees the "quest for self-fulfillment" as an aspect of the "democratization of personhood," a product of liberating historical forces drawn from the mainstream by marginal groups seeking entitlement, in a process that repeats itself as each new group begins the quest.[7]

Peale's positive thinking message, carried on a populist tide and undeniably middlebrow, was ideally positioned for a challenge from the highbrow critics in the seminaries and universities who redefined Peale himself into a kind of caricature. But when cut loose from the distortions of the controversy, Pealeism emerges in a new light, as a harbinger of the cultural diversity that has come to distinguish postwar America, religiously and otherwise. Peale not only challenged the bureaucratic power of the religious establishment but he also offered an alternative reading of its liberal theology. Mindful, as Lawrence Levine observed, that "trickle down" was no more operable in the realm of culture than of economics, Peale critiqued the sophisticated theological liberalism of mainstream churches and countered it with his version of practical Christianity, which accepted as its ministry the crises of everyday life.[8]

Although the bitterest battles between Peale and his critics were fought during the fifties, the debate ramified over the years, eventually caught up in the intensely argued issue of the very nature of American culture itself: whether it is becoming increasingly narcissistic and self-absorbed, or continuing the "pursuit of happiness" and political democratization. Peale's adversaries over the years have argued that there are fundamental flaws in his approach, many of which contribute to the disease of cultural narcissism. They point to what they perceive as the sin-denying quality of Pealeism, which makes the individual the center of the universe: It also camouflages personal denial, they suggest, by confusing success using simple techniques with underlying complex problems. Opponents of New Age religions saw it as another weed in that garden of religious heterodoxy.

An examination of the corpus of Peale's work reveals that the defining element in his thought is the metaphysical notion of positive thinking, long a culturally marginal outlook in American religious history.[9] It is based on the belief that through the mind and subconscious, utilizing techniques of positive thinking and affirmative prayer, one can achieve spiritual harmony and personal power. Yet Peale resisted any comparison with Christian Science or the mental science tradition. Once asked the stock question about possible influences on his work, he tapped a finger on his shiny desk for emphasis and stated forthrightly, "I belong to this," the solidity of the desk presumably symbolizing his institutional ties to Methodism and the Reformed Church in America.[10] Peale's eagerness to distance himself from Christian Science in particular reflected most mainstream religious opinion about a denomination described as radically idealistic, strongly antimedical, and savoring of the occult.[11] It was also the stance adopted by adherents of New Thought, a near-relative of Christian Science that had been engaged in an ongoing family feud with the Scientists and to which Pealeism had an obvious affinity. His response would have been congenial to his evangelical supporters.

A study of Peale's life, therefore, provides fresh evidence for examining two important contemporary considerations. One is the opportunity it affords to explore uncharted territory of the New Right, to rethink the old postwar liberal paradigm of the Right, which consigned all conservatives to the fanatical fringe of neo-Fascism.. The second possibility is to view Peale's career as a bellwether for major cultural realignments in the twentieth century, his priorities more symptom than cause of great subterranean shifts at work reconstituting the social landscape.

As Michael Kazin argued in the *American Historical Review,* scholars have just recently demonstrated a willingness to question their old characterizations of conservatism, and to apply the same intellectual rigor to studies of the Right that they have imposed on the liberal Left.[12] The liberal centrist image of the Right as an irrational neo-Fascist threat to intellectual inquiry and free institutions has been enduring and has discouraged historians and others from serious examination of modern conservative movements. Liberals have demonstrated that they are quite as capable of name-calling as their conservative counterparts. When Peale's critics described the affable preacher as a dangerous, heretical, neo-Fascist popularizer, they were giving vent to their own worst fears. There were, in fact, socially dangerous, intolerant right-wing activists who tried to exploit Peale's popular appeal, and he sometimes used poor judgment in his associations with them. It obscures an important part of the recent past, however, to throw a blanket indictment over his entire movement because of that. Revisionist interpretations of our past surely require moral sensitivity, but also a careful reading of the evidence.

In addition, the outline of Peale's life traces a significant cultural pattern, one with which we are still but limitedly familiar. A youthful denominational partisan, he subsequently placed his hopes in a generic, hegemonic Protestant "church," and when its seemingly liberal bias became too threatening, he abandoned it for his own grass-roots movement. Until about 1930 Peale was a traditional Methodist, schooled in the tenets of liberal theology and tried in the fires of Prohibition. When Prohibition failed, he kept his political energies alive through the assault on the New Deal, a grass-roots struggle that gained new strength from an alliance with a conservative religion—one that combined patriotism, laissez faire, and old-fashioned family values with a familiar evangelical Protestantism. Obliged to forsake his Methodist affiliation for a home in the Reformed Church in America at the time he went to Marble Collegiate Church, Peale came to show less regard for denominationalism and more for a hegemonic tribal "church." He believed that a revitalized church would shore up the foundations of Protestantism and restore old values. Lining up with conservative political lobbyists during the 1940s, Peale then added anticommunism to his political agenda.

At the same time, his religious message, increasingly critical of liberal leaders in the mainstream church, assumed a populistic, leveling tone. Its defining quality derived from the ancient folk belief in divine immanence, with the individual viewed as a participant in the spiritual energy suffusing the cosmos. Noncreedal, syncretic, and pragmatic, Peale's message had obvious appeal, drawing as it did from three streams of influence. It gained its symbolic language—about God, sin, and redemption, for instance—from the "great" tradition of the mainstream liberal church. Its emotional appeal came from its political priorities and its evangelical style. And its mass base was an adaptation of the ancient folk belief that empowered individuals with Nature's spiritual energy. It was this combination of conservative politics and popular religion that sustained the religious revival of the 1950s. But as religious energies splintered in the sixties, and Peale's own political expectations foundered during the campaign of John F. Kennedy, his lingering disenchantment with the mainstream church became manifest. Politically he threw in his lot with the National Association of Evangelicals, while religiously he became the Pied Piper of a dynamic grass-roots movement. He no longer pinned his hopes for the future on the "church," but rather on a "movement," which he himself was helping to shape.

The modern history of positive thinking offers useful social commentary. Understood as a religious sectarian movement, its central beliefs were at odds with orthodox Christianity, although it appealed to traditional Protestants, and with its naturalistic imagery had a special interest for women. Historically it has been quietly supportive of moderate social reform, but essentially apolitical. While remaining the exclusive property of popular culture, positive thinking can now be perceived in more professionally groomed models in the techniques of psychiatrists and psychologists, as well as pastoral counselors. When it threatened serious disruptions in the culture, it was embarrassed by skeptics who traced a dark path to mesmerism and mental manipulation. In the early years of the twentieth century, it finally possessed a sympathetic promoter in the religious mainstream. Its proponent was the multitalented Episcopal priest, the Reverend Dr. Elwood Worcester of Boston, who with his one-time patient and later colleague, the Reverend Anton Boisen, modified and adapted positive thinking to the clinical practice of pastoral care. Positive thinking, in part because of its association with metaphysical beliefs and a Worcester nexus to harmonial religion, found a niche in the holistic purview of psychosomatic medicine, part of the basic theoretical framework of clinical pastoral education.[13] Institutionalized and streamlined, and stripped of the controversy that surrounded Peale, it found its way into mainstream sermons and church counseling centers.

Peale has acknowledged his indebtedness to the tradition of modern metaphysical spirituality, although he utilized his sources selectively from Emerson on. In company with many other avowedly Christian religious scientists, he has understandably omitted reference to ancient myths of harmonial bonding. Earthy and naturalistic, the central myth of the tradition has emphasized oneness. Anthropologists and folklorists have suggested that the harmonial image is best recapitulated in the universalistic symbolism of union, as in the loving merger of mother and infant. An image of blissful, fulfilling organic oneness, it sustains a powerful and attractive spiritual sense of wholeness. Although the language of the myth is absent in most expressions of American harmonial religion, its symbolic presence helps explain the visible role of women in these movements as teachers and healers. In the mid-nineteenth century, the combination of metaphysical beliefs with nonheroic forms of treating disease, including forms of suggestion and mind cure, led to the emergence of alternate, irregular healing practices that proved especially appealing to middle-class women. One of the early scholarly investigators of the field termed the mind cure-positive thinking tradition "A religious medicine of feminine genesis."[14]

Women have been the overwhelming majority in all metaphysical groups in modern America.[15] Not only did they fill visible pastoral roles at a time when public ministry was denied to them but they were also in the majority as supporters and patients, crowding Elwood Worcester's waiting rooms and dominating Norman Vincent Peale's subscription lists. Women were attracted to positive thinking, or mental suggestion, for a number of reasons: It was a culturally acceptable response to life's physical and emotional difficulties; it afforded them leadership and entitlement in a way that traditional churches and traditional medicine did not; and it drew from an ideological orientation that seemed less authoritarian, autocratic, and elitist than most of the alternate arguments they heard.

One of the factors that turned the Emersonian-Swedenborgian message of harmonial spirituality to the cause of healing was the addition in mid-nineteenth century of the theories of Anton Mesmer. If Swedenborg provided metaphysical religion with its realm of mystical piety and Emerson its cultural references, it was Mesmer who contributed the revolutionary concepts about how the mind worked and how it might be approached. And it is in part because Mesmer has commonly been understood as a shadowy figure of the occult rather than as a modern contributor to theories of the unconscious that healing revivals like Worcester's and Peale's tend to draw down bitter charges of heresy.

The impact of perceptions of Mesmer's views was revealed in the decisive quarrel between two of the nineteenth century's better-known metaphysical groups, Christian Science and New Thought, when they contended over the extent to which a discredited form of mesmerism was necessary to the healing process. Intimately tied to this question of mesmerist influence was the problem of whom to acknowledge as the true modern founder of this mental healing movement, with the Scientists arguing for Mary Baker Eddy and New Thought advocates championing the cause of the self-taught Maine healer Phineas Quimby. For the later followers of Quimby in New Thought, he was the originator of positive suggestion and the innovative interpreter of the role of the mind in healing.

Conceived in the same regional nursery as Transcendentalism, mental healing in the nineteenth century developed out of a center in Boston, though it was not to remain confined there. Mrs. Eddy, New Thought teachers Julius and Annetta Dresser, and their related practitioners competed for adherents. One estimate speculated that at the time, that is, the last quarter of the nineteenth century, only one in five Bostonians with interests in mental healing chose to join Mrs. Eddy.[16] Along with other Eddy dissidents, the Dressers created the nucleus of a rather amorphous organization that in 1895 adopted the name New Thought to represent it as the "liberal wing of the therapeutic movement."[17] Of the many subgroups that made up the New Thought pantheon, the most distinctive was the Unity movement, or Unity School of Christianity, whose program has much in common with Peale's.

In the years between 1890 and 1910 when Christian Science and New Thought most firmly captured the public imagination, there was an audience particularly receptive to the ideas of mental healing because of the convergence of important historical forces. What contemporary G. Stanley Hall termed the "psychological moment," and later historian Nathan Hale described as the twin crises of "civilized morality" and the "somatic style," combined to nurture a pervasive American willingness to accept that the psychological, inner self held the clue to finding meaning in life.[18] Excerpts from the writings of William James, some commentary on Freud's work, sermons illustrative of liberal theology's emphasis on an immanent God, all found their way into the popular press. Romantic and idealistic, a rebuke to philosophical materialism as well as institutional neglect, the phenomenon of metaphysical mind cure held popular currency until the start of World War I. Healers were concerned not only with health but with creating examples of what Emerson and the Transcendentalists hoped would be a "living religion."

The most relevant modern model for the syncretic quality of Peale's ministry was Elwood Worcester. Worcester was the rector of Emmanuel Episcopal Church in Boston, whose healing clinic between 1905 and 1912 gathered the kind of media coverage that Peale's work would attract later. Worcester was an evangelical and a political liberal, one who interpreted the social gospel in terms of a ministry to individuals. A graduate of General Seminary in New York, Worcester went on to study for a Ph.D. in psychology at the University of Leipzig, where he took work with the famous Wilhelm Wundt. Although he had no quarrel with the social gospel message of his day, he believed that as a single emphasis of ministry it was incomplete because it was not "sufficiently spiritual" or "personal." Worcester respected the activist social program of the mainstream churches for its ability to help people "in the bulk, but it has no direct access to the depths of the individual conscience," he contended.[19] Worcester also feared that Christian Science and New Thought were drawing away members from the older churches.

With help from his psychologically trained clergy assistant and a few sympathetic local physicians—initially including the well-known neurologist James Jackson Putnam—Worcester opened his clinic in the church hall. His objective was to treat "functional" rather than somatic illnesses, or that range of maladies that has subsequently become the concern of behavioral medicine. His clinic was instantly crowded with patients, while Worcester himself was drawn into the national spotlight, invited to give talks nationwide to groups eager to replicate his healing experiment. Some commentators, picking up on a prediction made by William James, claimed that a religious revival was taking place, with Christian therapeutics as its defining quality.[20]

Worcester's celebrity quickly brought him the same kind of critical buffeting that Peale was to experience. Physicians were more censorious in their judgments than the clergy, although there were many vocal ministerial critics who regarded Worcester's version of "personal religion" as simply "Christian Science set to organ music." Press coverage of the movement and its critics undoubtedly facilitated the spread of Emmanuel models across the country, with clinics appearing in cities from New York to California. Fortunately for his reputation at the time, Worcester's developing interest in parapsychology, and his membership in the Society for Psychical Research, did not draw the attention of the media. Once the Emmanuel Movement began to fade from the national spotlight, Worcester turned more of his attention to these matters, where he could explore his fascination with such concerns as automatic writing and the evidence for immortality.

Worcester, like Peale, had a particular fondness for William James, whom he knew through his writings and as a fellow member of the Society for Psychical Research. What he found especially attractive in James's theories was his re-examination of the concept of the subliminal self, which was drawn from the work of the English investigator of parapsychology, Frederick Myers. For James, as for Myers, the subliminal self was comprised of an inferior function, which contributed to psychopathology, and a superior function, which produced a "subliminal uprush" of creativity, genius, and reflection.[21] The subliminal self was therefore the source of extraordinary psychic power, capable of yielding experiences inaccessible to rational consciousness and validated essentially by the experience itself. This phenomenological perspective, which James developed with exceeding richness in his 1902 *Varieties of Religious Experience*, made him appear as a partisan for the interior spiritual word of religion and metaphysics and fed his personal interest in parapsychology and psychic research. An apocryphal story about James and Worcester tells of a promise they made to each other: Whoever died first would communicate with the other through a code word. Some time after James's death in 1910, Worcester reportedly stopped in the middle of a conversation to say he had received the secret code, "pink pajamas."[22]

Despite, or perhaps because of, Worcester's interest in metaphysics, his healing work spread into other therapeutic areas, all revealing his commitment to a holistic approach uniting mind, body, and spirit. In his clinic alcoholics were eventually treated apart from the general patient population, in the associated Jacoby Club, where the techniques of total abstinence, group sessions, and a spiritual credo anticipated the work of Alcoholics Anonymous.[23]

In time Worcester's work crossed institutional barriers of religion and medicine that had previously been closed to him. His former patient and subsequent colleague, the Reverend Anton Boisen, was influenced by Worcester's model of holistic healing to initiate a program for seminarians in clinical pastoral training at Worcester State Hospital. Richard Cabot, a physician in Boston and a friend of Worcester's, also found ways to utilize his friend's ideas in his own work at Massachusetts General Hospital and in the community. And Helen Flanders Dunbar, who held advanced degrees in literature, theology, and medicine, and who was acquainted with Worcester, Cabot, and Boisen, served as the force behind the development of the Council for Clinical Training in Theological Education. Because of her multidisciplined education and practical experience, Dunbar developed a keen interest in the emergent field of psychosomatic medicine, an interest that was reflected in her work on the council and in her contributions to the medical field.[24]

Although there were many similarities in their approaches to Christian therapeutics, Worcester and Peale were personally quite different men. Worcester's high regard for Theodore Roosevelt could be seen in his regular engagements with rugged outdoor activities. More physically robust than Peale, he was also more socially engaged, moving in a wide circle of Boston professionals of the type Peale usually shunned. Too, he may have been less gifted as a preacher though more as a scholar than Peale. Yet both men attempted to bind marginal metaphysical beliefs to an evangelical, institutional mainstream through appeal to the power of in-dwelling Mind or spirit. Both enjoyed wide popular support and vigorous professional criticism.

But Worcester's name quickly faded from national attention once the movement disappeared from the press. Peale's has been preserved and sustained because it so thoroughly penetrated popular culture. The Phenomenon of Pealeism was a composite of many factors, but for adherents it was a claim of entitlement—to better health, better well-being, better relationships, and a better quality of life. And an important reason for the movement's success was the extent to which it recapitulated Peale's own life experiences.

Notes

1. The survey, conceived by George Gallup in consultation with Robert Wuthnow, found approximately 5 percent of the population involved with each group. Cited in Wuthnow, *The Restructuring of American Religion* (Princeton, NJ: Princeton University Press, 1988), pp. 120; 332, n50; 336, n8.

2. Lawrence W. Levine, *Highbrow Lowbrow: The Emergence of Cultural Hierarchy in America* (Cambridge, MA: Harvard University Press, 1988), p. 252.

3. This line of argument is developed in Peter Clecak's *America's Quest for the Ideal Self: Dissent and Fulfillment in the 60's and 70's* (New York: Oxford University Press, 1983). See pp. 151ff.

4. Christopher Lasch, *The Minimal Self: Psychic Survival in Troubled Times* (New

York: Norton, 1984). See also his *Culture of Narcissism: American Life in an Age of Diminishing Expectations* (New York: Norton, 1978).

5. Philip Rieff, *The Triumph of the Therapeutic* (New York: Harper & Row, 1966).

6. Warren I. Susman, "'Personality' and the Making of Twentieth-Century Culture," in Susman (ed.), *Culture as History: The Transformation of American Society in the Twentieth Century* (New York: Pantheon, 1973), pp. 271–85.

7. Clecak, *America's Quest,* pp. 22ff.

8. Levine, *Highbrow Lowbrow,* p. 226.

9. There is a useful discussion of metaphysical beliefs in the article by Robert Galbreath, "Explaining Modern Occultism," in Howard Kerr and Charles C. Crow, (eds.), *The Occult in America* (Chicago: University of Illinois Press, 1986), pp. 11–37.

10. Interview with Norman Vincent Peale, Pawling, New York, June 17, 1987.

11. R. Laurence Moore has said that as Mary Baker Eddy reinterpreted her sources of mystical inspiration, from people such as Phineas Quimby and Emmanuel Swedenborg, her views became associated by contemporaries with Theosophy and expressions of the occult. See Moore, "Mormonism, Christian Science, and Spiritualism," in Kerr and Crow, *The Occult in America,* pp. 143–49.

12. Michael Kazin, "The Grass-Roots Right: New Histories of U.S. Conservatism in the Twentieth Century," a review article, *American Historical Review* (97), February 1992, pp 136–55.

13. See Robert Powell, "Healing and Wholeness: Helen Flanders Dunbar and an Extra-Medical Origin of the American Psychosomatic Movement, 1906–36," doctoral dissertation, Duke University, 1974.

14. Donald Meyer, *The Positive Thinkers* (New York: Pantheon, 1980), p. 72.

15. While statistical evidence on gender division is lacking, there is abundant impressionistic commentary on the preponderance of women in the movement. Elwood Worcester, for example, kept records for his counseling clinic, and they indicated that the proportion of women to men was about five to one. The women tended to suffer from ailments related to neurasthenia, the men from alcoholism. An assessment of Worcester's files was made by Dr. Richard C. Cabot and appeared as "New Phases in the Relation of the Church to Health," *Outlook,* New York, February 20, 1908. Newspapers and journals confirmed this gender division in the Emmanuel Movement in articles on New Thought and other mental science groups.

16. Stephen Gottschalk, *The Emergence of Christian Science in American Life* (Berkeley: University of California Press, 1973), p. 116.

17. Ibid., p. 99; Horatio Dresser, *History of the New Thought Movement* (New York: Crowell, 1919), p. 187.

18. This is Hale's argument in his *Freud and the Americans* (New York: Oxford University Press, 1971).

19. Worcester commented on his work in a series of articles for the *Ladies Home Journal.* In February 1908 he wrote, "My experience in the ministry has revealed to me the immense number of sad, dispirited, unsettled men and women who haunt our churches, wistfully looking for the help which they seldom receive."

Two members of Worcester's family were especially helpful in providing me with reminiscences and historical resources. A collection of "Emmanuel Scrapbooks," contemporary newspaper and journal clippings kept by members of Worcester's family, was generously made available to me by Worcester's grandson, Carroll Worcester Brewster. I also had interviews with Worcester's daughter, Constance Worcester, on January 14 and 15, 1985, in Boston, and with Carroll Brewster on December 12, 1984, in Geneva, New York. The Episcopal Church Archives in Boston also contain Worcester material, which I used.

20. William B. Parker, "What Is It All About? *Woman's Home Companion*, March 1909. Parker was the editor of the shortlived journal *Psychotherapy*, which covered the Emmanuel Movement.

21. Henri Ellenberger, *Discovery of the Unconscious* (New York: Basic Books, 1970), p. 314.

22. The story was told to me in interviews with Constance Worcester (Boston, January 15, 1985) and Carroll Worcester Brewster (Geneva, December 12, 1984).

23. A work that continues in the post-World War II tradition critical of Peale is Wendy Kaminer, *I'm Dysfunctional—You're Dysfunctional: The Recovery Movement and Other Self-Help Fashions* (Reading, MA: Addison-Wesley, 1992). The book associates Peale with the mental healing techniques of Alcoholics Anonymous and other Twelve Step programs, an approach the author found "authoritarian and conformist" and political. See pages 6 and 151ff.

24. Robert Powell, "Healing and Wholeness," doctoral dissertation, Duke University, 1974.

I

The Man

What Sydney Ahlstrom descriptively termed "The Phenomenon of Pealeism" is an historic and cultural reality composed of three indivisible parts: the man, the movement, and the message. While the message has had the greatest transformative effect on American life, it could not have existed apart from the other two dimensions.

For what is especially clear on close examination of this phenomenon is the extent to which it was a manifestation of Peale the man. Not surprisingly in a populist movement like Pealeism, the public offering, the message, and the messenger existed in a kind of symbiotic relationship. Particularly notable is the way Peale was able to define the movement in terms that reflected his own personal needs and convictions.

The eldest son of a struggling Methodist preacher and his wife, Peale had strong emotional bonds to both family and denomination that inevitably revealed themselves in the movement. The pietistic evangelical Methodism that he learned in the bosom of his emotionally intense family provided the basic ideological structure around which he molded his message of metaphysical positive thinking. A resident of Manhattan for most of his adult life, he self-consciously preserved remnants of his Ohio background. A portion of the image was studied and politically useful, but essentially the core image was true. Devoted to his family and his roots, Peale imagined himself conserving their values, extending them through the tools of modern technology.

Yet he was also a complicated man whose celebrated reputation as a positive thinker belied the alternate side to his personality. What he understood initially, that the message he was developing was strongly autobiographical, he continued to include in his presentation, ritualizing it to become another means for developing bonds with his supporters. His gospel of personal religion, which placed greater reliance on individual experience than on institutional church life as a guide to belief, was also an important statement about Peale's own convictions regarding personal autonomy.

Peale was also a man with remarkable public talent. He could captivate almost any audience with his charming accessibility, good humor, and matchless style as a storyteller. Embroiled in controversy all his life, he was rarely criticized personally by his opponents, some of whom seemed at pains to demonstrate that he had been taken in by unscrupulous friends he unwittingly trusted. His personal talent, his life circumctances, his psychic tracings, all provided the formative stuff out of which the phenomenon developed. The gospel of positive thinking was richly autobiographical.

CHAPTER 1

A Thoroughly Methodist Beginning 1898–1921

> Men are what their mothers made them. . . . When each comes forth from his mother's womb, the gate of gifts closes behind them. Let him value his hands and feet, he has but one pair. So he has but one future, and that is already predetermined in his lobes and described in that little fatty face and pig-eye, and squat form. All the privilege and the legislation of the world cannot meddle or help to make a poet or a prince of him.
>
> History is the action and reaction of these two,—Nature and Thought; two boys pushing each other on the curbstone of the pavement. Everything is pusher or pushed; and matter and mind are in perpetual tilt and balance, so.
>
> Emerson
> Essay on Fate
> c. 1852

From the moment Norman was born, his parents were certain he was a Methodist even before they were sure of his name. But he was soon designated by name as well: By his middle name Norman Vincent Peale would be honoring Methodist Bishop John Vincent of the Ohio Conference, who, significantly for his namesake, introduced the original Chautauqua Assembly in western New York as a form of popular education. For a couple as genuinely serious about their denominational ties as Anna and Charles Clifford Peale, the dedication of their firstborn to the church was never a question. It is not difficult to imagine that even as she watched him in his cradle, Anna Peale saw another Methodist minister, possibly even a bishop, being formed.

In late-nineteenth-century small-town Ohio where the Peales lived, church membership was a primary form of identification. And since both Anna and Clifford were passionate and committed about everything they did, they were an intimate part of the invisible network that held church, community, and local families together. Norman in later life remembered his parents as "vital" and "lively," as being at the center of a busy household with many interests. The manner in which this new son arrived in the world was testimony to the emotional intensity his parents invested in almost all situations, even those far less momentous. The young couple had decided that Clifford would guide the drug-less home delivery of their baby himself.

A Methodist minister at the time, Clifford had earlier trained as a doctor, and he felt confident that he could demonstrate the harmonious cooperation of science and religion while not incidentally saving the cost of a hired professional or a hospital setting. With Anna managing the labor and the young preacher serving as midwife, the pair successfully guided the very unconventional introduction of tiny Norman to life in Bowersville, Ohio, on the sunny spring morning of May 31, 1898. The private drama for the small family marked a turning point in Clifford's medical career: This would be the last formal, medically related procedure he would perform; two sons born subsequently to the couple, and a set of twins who died in infancy, were all delivered by other, licensed physicians.

The decision to bar outsiders from the intimate event said a great deal about the young couple's sense of family bonding. Even by nineteenth-century middle-class standards, this was an unusually private affair, an extraordinary instance of family togetherness fraught with memories. It set up the image of the Peale household as a small, tight haven. Not surprisingly, however, it was Anna, responsible for making the bold decision to forego modesty for a home delivery, who would often shatter the shell of the family cocoon, entering the larger world and bringing its treasures into the home. Clifford's role in this most primary rite of passage would be repeated symbolically many times in the future as he orchestrated other major transitional events in his son's life: his conversion, his entry into the church, his marriage, and his reception into the ministry.

Always larger than life to his eldest son, Clifford Peale was temperamentally suited to be a midwestern Methodist minister. He was a gregarious, voluble man, guided to the ministry by some of the same reasons that had earlier led him into medicine: He enjoyed both being with people and the feeling that he was doing them some good. As important was the fact that both professions involved him with life in the "real" world and not the sedentary, bookish pursuits of scholars and writers. An activist by nature, he had the sort of unpretentious appearance his son would come to share: He had a modest stature and at the time an average build, which betrayed no signs of the pounds the years would add. Bowersville was only his second assignment on a Methodist circuit, and despite the meager material compensation, he looked to the future more expectantly than he had as an assistant city health commissioner. But the future inevitably held surprises, and an important one for father and son was that Norman would eclipse his father's career by the time he was twenty-seven and Bishop Vincent's by the time he was in his mid-thirties.

Practical Methodists: Clifford and Anna

When Clifford Peale died in 1955, it was said of him that he had three "all-consuming" devotions: to God, to his family, and to his country.[1] It was a fitting eulogy to an evangelical Methodist, a Republican, and the father of three professionally "successful" sons—two in the ministry and one in medicine. The Puritan work ethic, to strive and "make something of yourself," was as much a part of his mental makeup as his Methodist sense of grace, free will, and the need to redeem the nation, one person at a time. The fight to save the nation from liquor was his as well, and as a lifelong Republican, he championed the fight of the Methodist Church and the Anti-Saloon League for prohibition. He was supportive of the moral code of the Methodists, and though he was known to smoke an occasional contraband cigar, he was privately and publicly a model of Methodist propriety.

Twenty-eight years old when Norman arrived, Clifford had been born in Lynchburg, Ohio, on August 4, 1870 to Laura Fulton Peale and Samuel Peale, his father the proprietor of a local store. Through his mother he was related to inventor Robert Fulton, and on his father's side he traced a distant ancestry to artists Charles Wilson Peale and Rembrandt Peale. In his youth Methodism had already caught fire in the transmontaine Midwest through the efforts of earnest, largely untutored, circuit riders, and he was able to recall in later years the thrill of attending revivals and even an occasional camp meeting. After high school in Lynchburg, he enrolled in Ohio Medical College in Cincinnati. He graduated in April 1892 and was qualified by the standards of the time for a career in medicine, despite his lack of a university degree.

The possibility of becoming a preacher had entered his mind many times, but feeling he lacked the "call," he dismissed it and moved to Milwaukee to work as assistant health commissioner. Alone and lonely in the strange city, he attended the Summerfield Methodist Church, became an usher there on Sundays, and spent the rest of the week working long hours to help combat a serious epidemic of smallpox. Once while listening to a sermon at the church, he was convinced he heard a voice beside him say, "That is what you should do." Reporting direct contact with the realm of the spiritual was not an uncommon experience for Clifford or the other members of his family, eventually including Norman as well. Like other Methodists of their day, they felt no embarrassment about testifying to moments of mystical and pietistic encounters, ecstatic moments filled with sounds and visions. But the smallpox epidemic had prior claim on Clifford's attention, his time spent offering vaccines and a great deal of persuasion to patients fearful of the prospect of injection. Eventually his own health suffered, and he returned to Lynchburg for surgery. In the early postoperative period, when his recovery seemed questionable, his mother "promised" him to the Lord if divine healing spared her son's life. His mother's commitment seemed to coincide with his own sense of a genuine "call." Restored to health, he closed his medical office and was ordained to the Methodist ministry in 1895 at the Cincinnati Conference. It was the same year in which he married Anna DeLaney.

His "call" was Clifford's primary qualification for the ministry, his medical education representing more formal preparation for the role than most local preachers possessed who applied for ordination. His theology was essentially a localistic product, derived from the tradition of evangelical morality and the theological duststorms that swirled in Gilded Age America. The furor in theology had been precipitated by Darwinian evolutionism and higher criticism and had been gathering strength over the years in conflicts between liberals and conservatives. By midcentury it had essentially become a contest between those who wanted to synthesize the old and the new in a "progressive orthodoxy" and others who favored rejecting creedally based theology for a form of religious intuitionism.[2] While clerical leaders and seminary professors attempted to chart a path through the confusion, people in the pews, according to historian Henry May, were religiously stranded on a "summit of complacency."[3] To outsiders, the mainstream church looked healthy, even vital, while to insiders it seemed as if the institution suffered from a crisis of morale and excessive formalism.

Responses to this perception of internal apathy got worked out in a number of ways. There was evidence of the growth of metaphysical groups, with their healing message. Pulpit princes like Phillips Brooks and Henry Ward Beecher extended their efforts to reach the unchurched urban residents. And most relevant for Clifford Peale was the struggle within Methodism to suppress the premillennial perfectionism of those caught up in the Holiness Revival without killing the spirit of renewal. As a person whose medical training had given him a respect for science, Peale was repelled by the emotional extravagance of perfectionist revivalism and was more in sympathy with the evangelism, though not the literal premillennialism, of Dwight L. Moody, with its measure of muscular Christianity.[4] Science aside, however, Peale was a religious sentimentalist who understood the true test of faith as a personal relationship with Jesus Christ. It was a view with which both the liberal Beecher and the conservative Moody were also in accord. His theology came not from books but from experience—his own and his parishioners—and like a growing organism found its vitality in the environment. No systematic corpus of beliefs defined his religious thinking; no inkling of metaphysical spirituality intruded on his ministry, although it was the case that mid-western Methodist perfectionism, emotional and intuitional, could appear as a rural, less polished relative of the New England model of metaphysical piety.

The experience of nineteenth-century evangelical Protestantism was, next to his family, the most decisive influence in Clifford Peale's life, and he would pass it on to his eldest son virtually intact. Wedded to the secular culture through the civil religion defined by the Anglo-Protestant establishment, his evangelicalism was the expression of nonurban mainstream Methodism. Its sources came more from experience and the culture than from the academy: The Bible, to be sure, but rivaling it in importance were the lessons taught by the McGuffey Reader and Horatio Alger stories. His Wesleyan "heart religion" persuaded him of the need for an "inner" encounter with the spiritual presence, which placed a strong emphasis on the role of free will and human agency. He admired the so-called self-made men of his time, who through the gospel of work demonstrated the effectiveness of the capitalist system: The successful entrepreneur was presumably a testimony not only to financial acumen but to moral virtue as well; he was a person who could be trusted.[5] In the senior Peale, his evangelical Methodism readily blended with the Protestant civil religion identified with the Republican Party, as it did for most of his evangelical contemporaries. With minimal adjustments, this tradition—conservative, evangelical, localistic—would function in the same way for his son.

In Peale's "heart religion" emotion was courted rather than avoided, and sentimentality ran strong. He was more receptive to the "old time gospel" of Moody than to the urban liberalism of Beecher, although Beecher remained one of his personal favorites. Throughout his life, Clifford loved to sing and hear the gospel songs popularized by Moody's soloist, Ira Sankey. Tearful and whole-souled, he could range through many verses himself without a text. And his own evangelistic efforts indicate that he acted on Moody's observation that "It doesn't matter how you get a man to God provided you get him there": Results were what counted.[6] His religion, like his music, was intense and robust—the source of meaning for his life. His earlier scientific training led him to part company with those among the evangelicals who would seem his natural allies—he had difficulty, for example, with Moody's views on evolution—but on the core issues that interpreted religion to the culture—Protestant morality, individualism, hard work, thrift, self-help—he was in harmony with the celebrated evangelist.[7] Horatio Alger's winning models were his as well.

But whereas Moody pictured life as steady warfare against the devil, and marshaled congregations with the rousing music of "Hold the fort, for I am coming," Peale's idiosyncratic postmillennialism allowed him to see a more corporeal foe. Rum, wet, liquor, alcohol, all were simply different words for the same enemy. For virtually all Methodists it meant accepting Wesley's injunction against the use of spiritous liquors. By the latter half of the nineteenth century, however, firmly a part of the Protestant civil religion, Prohibition had come to mean something more: No longer applicable only to personal conduct, it became an organized battle against saloons and sales of alcoholic beverages and for a national referendum on alcohol.

The temperance crusade, viewing urban immigrant groups as the heaviest drinkers, came to symbolize a cultural struggle between native-born Protestant rural residents and immigrant, Roman Catholic, urban dwellers. And

Ohio was the heartland of Prohibitionist sentiment, the organizing home for the Prohibition Party in 1869, the Women's Christian Temperance Union in 1873, and the Anti-Saloon League in 1893.[8] From the early failure of the political activism of the Prohibition Party, its descendants learned the value of engendering grass-roots support, enlisting the efforts of local pastors, church leaders, and civic groups. For evangelical Protestants like Clifford Peale, the Prohibitionist movement became as much a religious as a social-political crusade. And Clifford's wife, but especially his eldest son, came to see it as their fight too, with Norman cutting his political teeth on the literature of Prohibitionism. It was a form of grass-roots activism with telling consequences for the young man's future.

Had ideology rather than emotion determined Clifford's marital choice, he might never have considered Anna DeLaney for his bride: Her father had been born a Roman Catholic, and Irish to boot. Clifford was said to have been instantly attracted to the young woman with blue eyes and fair hair, and his feelings remained the same on learning her father had been reared as a Catholic in Ireland. Anna herself was, of course, a Methodist, born in Paris, Kentucky, on February 7, 1875. Her father, Andrew, reared in Dublin by two wealthy aunts, had got himself to Ohio by stowing away on a ship. He landed a job with Potts's cooperage business and eventually married the owner's daughter, Margaret Potts. Margaret was described as a sedate woman with an "English" background, temperamentally the reserved counterpart to her more gregarious husband. Of Anna's personality it was said to be more like her father's than her mother's, although her eldest son remembered her as having just the right combination of reserved British bearing and gay Irish lyricism. He held treasured recollections of a woman who combined the ability to recreate images of spirits and fairies in her Irish stories with a serious Methodist commitment to save the world through foreign missions. She was as devoted a Methodist as Clifford, and their marriage was consecrated in the Methodist Church.

Shortly after their wedding, Anna and Clifford Peale drove a horse-drawn wagon to the new preacher's first charge in Sugar Tree, in southern Ohio.[9] Just over a year later, in 1896, they moved to Bowersville, where Norman was born and where Clifford served a circuit of three churches. Historians have described the gradual emergence at the turn of the century of a new professional class, and a bureaucratic middle class beginning to enjoy better incomes and social status—some who would come to patronize Norman's remedies for anxiety—but that change had yet to affect most small Ohio towns. Clifford Peale's parishioners were farmers, shopkeepers, and teachers, their singular social status made more noticeable by the appearance of an occasional banker. His small salary almost always in arrears, Clifford was often paid in kind from the farmers' fields and the housewives' kitchens.

In 1900, the Methodist Conference sent Clifford Peale and his family to the small town of Highland, seventeen miles from Bowersville. They stayed for three years, during which time a second son, Robert, was born on January 2, 1901, and Norman became an older brother. He dutifully took on the responsibilities of the eldest and retained them throughout a lifetime of concern for his younger siblings. His first reaction to the squawling reality that would now push him aside to claim an important share of his parents' attention was not so generous, however, and he verbalized the wish experienced by many older children on such occasions: "Dead the baby, dead the baby."[10] As evidence of his early sense of responsibility, he told the story later in life about the Sunday morning when, as his father was preaching, the baby's restlessness prompted his mother to send young Norman home for crackers; on his return, he carefully dragged an oversized five-pound sack down the church aisle. From Highland, too, he carried the "traumatic" memory of seriously injuring a little neighbor girl, severing her finger with a lawnmower after she inadvertently stuck it in the blades. But there was a successful, even heroic, outcome as his father secured the finger in place until the girl could be taken to a hospital and treated.[11] As "traumatic" as the accident was to Norman (and to the girl), of equal importance to him in his recollection of the event was his father's salvific role, which he carefully stored in his personal psychological record.

Clifford Peale was reassigned again in 1903, this time to Asbury Church in the urban setting of Cincinnati. By the time Norman was ready to leave for college, he had lived in seven different Ohio locations and more than a dozen Methodist parsonages.

Every September, at "Conference time," the presiding elder, or district superintendent, announced the clergy assignments for the coming year, with the average stay in one location three years. The system had the advantage of instilling new life and ideas in local congregations. It also limited the potential for personality-centered, clerically dominated parishes while also creating the impression of a democratic rotation of ministers among "good" and "bad" charges. Ohio Methodist Church members were particularly eager to secure and protect active lay participation in the life and politics of the church. But as John Wesley himself appreciated when he instituted the circuit system utilizing only unmarried men, it was a method of evangelizing not conducive to conventional family life. Even with nonitinerant, "settled" ministers, the regular rotation every few years meant that it was the clergy families who suffered; wives were forced to improvise furnishings and limited creature comforts in yet another old parsonage, and children had to make new friends and attend new schools, usually after the school year had started.

Some children, adjusting to the inevitable, probably thrived on the adventure of relocating; others, wearied of making introductions and starting over, created their own sense of stability within themselves and their families, and to a lesser extent, within their fathers' churches. Norman fell into the latter group; shy, skinny, sensitive, with a self-described "sense of inferiority," he found his emotional pleasure in his home and family. The steady moving and the social estrangement it produced had lifelong consequences. He always thought of himself as an outsider, had few friends, and was continually ill at ease with strangers or in small groups. It may have contributed, too, to his adult identification of himself as a "nomad," an observation made when he was in his eighties and still logging 200,000 airline miles annually, still in constant motion toward another audience in another distant city.

Parsonage life was also a daily struggle for the family financially. Methodist churches, appealing more to a solid or aspiring middle class than to an upper class, were not noted for paying high stipends to clergy, and Clifford Peale's was no exception. His salary, in addition to being late, was also small. Anxiety over grocery money was constant, furniture was someone else's discard, and new clothing was an event. It had not escaped Norman's attention that the inevitable banker in every small church, enjoying a new car and the visible accouterments of a higher life style, knew that when tragedy struck his family, at whatever the day or hour, he could count on the underpaid preacher to be available for solace and comfort. When those situations arose, Clifford reminded his son not to equate service with compensation. But the experience taught strong lessons. Poverty—and perhaps even more so near-poverty—was a stern taskmaster, a grim but real prospect in a Puritan-based culture that equated it with sloth, heredity, and lack of diligence. Dwight L. Moody speculated often about the connection between poverty and personal sin. The legacy of always living on the edge drove the adult Norman Peale to Promethean efforts for capital accumulation, into the company of successful businessmen, whose self-made financial independence mirrored his own need for economic and status security.

The Peale family left Asbury Church in 1905 and moved three more times before Norman left to attend Ohio Wesleyan University in 1916. They were at Grace Church, Norwood, between 1905 and 1910; at First Church, Greenville, from 1910 to 1913; and then at First Church, Bellefontaine, from 1913 to 1917. Although many large mainstream churches were struggling at the time to respond to the demands of either the social gospel appeal or the Emmanuel healing revival inspired by Elwood Worcester, neither movement made an obvious intrusion on Clifford Peale's modest congregations. In Methodist churches in small towns in Ohio, the social gospel was translated as Prohibitionism, and the most noticeable product of a national coming of age was the proliferation of consumer goods. Alcoholism, and neurasthenia when it was diagnosed, were still described as essentially moral and character problems by most rural, midwestern Methodists, although eclectic, metaphysical healers were making some notable converts even in Ohio.

Its roots fixed in the nineteenth-century bourgeois ethos, the Peale family nurtured its emotional bonds as repeated moves threatened always to unsettle it. Home life was the focal point of the family, but because a small-town ministry was more like a mom-and-pop store, there was little distinction between personal life and work, with the church expecting everyone to be a recruit. Anna Peale was the emotional center of the family and its financial manager and Clifford the acknowledged ceremonial, patriarchal head, whose social gregariousness drew him into such a wide range of activities that his young wife was left with the major responsibility for home and family. Norman's memories of these early years are anecdotal, capturing snapshots with a narrow-lens mental camera focused only on the smiling participants in the foreground. His pictures of his father and himself are vivid and distinct, those of his mother fewer, softer, ambiguous, and elusive.[12]

Childhood Revisited

In the early years of this century, most ministers could still trade on the nineteenth-century idealized image of their calling, with its assurance of status and social recognition. It was and remained Norman's view of his father. For him, his father was a "he-man" sort of preacher, a muscular Christian able to convert drunks, a dying prostitute, and a whole gamut of backsliders into a belief in saving grace. Clifford enjoyed his ministry, never regretted his career change, and relished the challenge of bringing the most wayward sheep into the fold.

In his reminiscences, the grown son remembered a faultless father and a dedicated minister, recollections ultimately more revealing about the son than the father. Norman recalled, for example, his father's ability to persuade the town drunk to accept the minister's altar call and a commitment to sobriety, and of a similar success story with an alcoholic clergy colleague. When Norman was a boy of nine or ten, as he remembered it, his father took him on a pastoral call to a fatally ill prostitute—to protect his father's reputation and to expose the son to the "evil of this world" he was told—and overhearing the girl's confession, he found himself experiencing "the wonder and glory of the ministry, and the majesty and power of the pastor."[13] As the eldest child, Norman often tagged along on pastoral visits, and, young and impressionable, he collected strong memories of the level of congregational affection his father enjoyed.

Unlike his mother, who could be short-tempered and outspoken, Norman's father was an easygoing person, a man who was usually straightforward in his interactions and who found it natural to express his feelings. Tears and weeping were indications of spiritual stirrings rather than signs of delicacy. Clifford taught his children their moral lessons casually and incidentally. Norman remembered the time his father pointed out that it was a Christian obligation to give one of their scarce dollars to a panhandler at Christmas, and that it was wrong for neighborhood children to call a grimy streetcar worker "Greasy Dick." Some lessons required sterner, more physical measures, but revealingly, Norman never told the reasons for such chastisement, and referred to a "licking" he received only in passing.

The anecdotal images in Norman's family scrapbook lose some of their gloss when attempting to capture weighty events like rites of passage. Almost predictably, they are moments from adolescence. There was the occasion when he was once again in his father's company on a pastoral visit. They were living at the time in Greenville, and the two went on foot to see a family in the countryside. On the trip back to town, Norman began unfolding the dimensions of his social problems, including his terrible shyness. His father diagnosed the symptoms as evidence of an "inferiority complex," but there was to be no medical attention because it was unavailable locally and too expensive in any case. What Norman needed, said his father, was to make a total commitment to Jesus Christ, because Jesus was the best doctor, who could free his mind and change his life.

The latter-day positive thinker identified this advice as an invitation to commitment. He was "profoundly impressed" by his father's words, he remembered, and as the two knelt in prayer, Norman acknowledged that he was giving his life to Jesus and letting go all of his inferiority feelings. Norman never isolated a singular occasion as his conversion experience, although he regularly alluded to being converted at one of his father's revival meetings. The clarity of this recollection recommends it as a particularly meaningful event. He would refer to it often in his later ministry. Curiously, by his own admission, this ceremonial rite had no effect on his inferiority complex, which remained unchanged until his college years, when he claimed to learn finally how to "lick it." In fact, he said, as with a person recovering from alcoholism, it was a problem that never really left him.

A more vivid picture emerged from a father and son discussion about sex. He was having "bad thoughts," he told his mother, and he wanted to talk with "Papa." When his father came to his room, Norman told him about the bad thoughts and the tossing in his bed at night. When asked if the thoughts were about girls, Norman said yes. "As your sovereign self, disown the thoughts. Don't act on them," said his father. The thoughts were like "Indian scalps. You have to scalp them and hang them on your belt." This was strong Methodist medicine about the value of self-control, specifically the ability of the mind and the will to control the passions, including the libido. It may also have been incorporated with oedipal fears, later manifest in the younger Peale's strongly ambivalent feelings about authority figures.[14]

Most of the scenes his memory framed were more joyous. He remembered the financial luxury of going to an amusement park with his brother Bob, and of their going together to see their new sibling Leonard, born April 20, 1912, who happily survived the predictions about his fragile health. On one Christmas, the two older brothers shared the excitement of a new, actually secondhand, bicycle, a gift dearly prized because they knew of their parents' scrimping to buy it. And Norman remembered the time when, thinking he was alone in his father's church, he went into the pulpit to imitate him and to practice his own oratorical skills. His father, in the shadows, overheard him and praised him for his talent, observing that Norman had the ability to surpass him if he took himself "out of the picture" and put Jesus in the center. His father was concerned that Norman's preaching had too much "self" in it.

His stories tell of a boy's highly idealized image of his father, although his real emotional bonding was with his mother. In a psychological sense, his idealized vision of his father may have represented a form of the ego ideal, an aspect of the redirected narcissism of childhood.[15] His mother, however, provided the emotional ballast in his life. He knew from childhood that his "mother always wanted me to become a preacher."[16] Anna Peale herself was an active participant in religious gatherings, and had she lived in a different time, she might have considered the ministry for herself, though politics or other forms of public office would probably have interested her more, her son later guessed. He developed an interest in political life himself.

People frequently told Peale that he resembled his mother, not only in appearance but in personality. Anna Peale was a bright woman, an avid reader gifted with a photographic memory, which she passed on to her son, and which, like Elwood Worcester's claim to having total aural recall, Norman regarded as one of those "special" gifts holding metaphysical connotations. Whenever she spoke in public, it was always without notes, a technique her husband also used, as would her son in later life. He described her as an accomplished speaker and a storyteller, skilled as a "communicator." Her reading interests inclined toward poetry and literature, enlivened by reports she collected from the foreign mission field and the strangely new and exotic cultures they described. Spirited and independent, she cast her first vote in 1920 for a Democrat, as her eldest son, also casting his first vote, followed the "Peale family men's tradition" and voted for Harding, the Republican. Years later after Norman Peale had settled in as the minister of Marble Collegiate Church, he arranged to take his whole family to Ireland so that his mother could trace her family lineage.

Although as a Methodist preacher's wife she had an assumed identity, Anna Peale discovered ways to apply the role to situations that gave her pleasure and had social utility. Often the two were synonymous. As her children grew and became more self-sufficient, she was able to channel her abilities to activities outside her husband's local parish: to a seat on the Findlay, Ohio, Board of Education; to leadership positions in various Ohio women's clubs; and to a position on the Methodist National Executive Board of the Women's Foreign Missionary Society. As a result of the denominational assignment, she was invited on an eight-month visit to mission stations in the Orient in 1934, followed by a full round of speaking engagements dealing with missionary activity when she returned. For some of these lecturing assignments she was given modest honoraria, welcome supplements to the family income, for although by the 1930s her sons were grown, Clifford's income remained small and uncertain. With advancing years, she began to weary of the exactions of life in the parsonage. Although she apparently never mentioned it to Norman herself, Anna felt frustrated by the inability to afford a decent car or the travel costs of attending meetings.[17] And often anxious and not well, she was not a particularly effective model of positive thinking.

These extradomestic opportunities became available to her, as to most other women, only after the demands of child-rearing had passed. The woman of Peale's boyhood memory was domestic, typically "feminine," and devout, a busy, caring, well-organized homemaker who found time to make her own Christmas candy, who made her children feel that she was totally devoted to them. The other image of his mother, as a mature political activist, independent world traveler, and accomplished speaker, was superimposed on the earlier one. In a sense, until her declining years, she seemed to combine the best of both male and female qualities.

Like the earnest Methodist parents they were, Anna and Clifford Peale helped their eldest son understand from childhood that he must "make something of himself." When Norman finally decided for the ministry, his mother experienced the pleasure of one who's fondest dream had come true. Many years after his first public sermon, a family acquaintance wrote to Norman to describe his mother's reaction as the two friends listened to his initial effort. His mother, said the friend, was overcome with emotion, great tears rolling down her cheeks, which she allowed to fall on her new dress unchecked so as not to call attention to herself. When Norman had concluded, Anna Peale realized that the tear stains had ruined her dress, but she allowed that despite the costly casualty, it was a small price for such a "wonderful experience."[18]

Her affection for her eldest son was returned in full. He took as much pride in her accomplishments as she did in his. Once he was appointed to his first full-time pastoral charge in Brooklyn, New York, he began inviting her to speak from his pulpit about her experiences—whether foreign travel or her work with the Methodist denomination. As his parents aged, he took care of their financial and other personal needs, even guiding his father, after many years as a widower, out of a romance the son regarded as questionable.

Some years after his mother died in 1939, Norman wrote to a friend from his early Ohio days:

> It would be quite difficult, even impossible, for me to describe the influence my mother had on my life. In addition to the usual influence which a fine Christian mother has on her son, I may say that I was always impressed by her eager interest in life. Everything for her had charm, romance, color, beauty, and I marvelled at this, and believe that among the many gifts which she gave to me is this insatiable and never-ending interest in life.[19]

He found it comforting at the time of her funeral to recall her irrepressible humor and spirit, and sometime later added retrospectively, "Indeed, for me she never really died."[20]

Peale's metaphysical understanding of life after death, which came to assume an ever-larger place in his thinking after his ordination to the ministry, was especially confident when it involved his immediate family. In the years after his mother's death he had a number of clairvoyant experiences, during which he saw and heard her, as well as his father after he had died, in episodes he always found warm and reassuring. In fact, his ties to both parents were so strong that he was convinced they would never be broken, even by death. Four years before his father's death, Norman wrote to him, "I have no doubt at all about the reality of the invisible world.. . . . I want you to know how much I love you. Within two days now I will be fifty-three years old and I have loved you for those fifty-three years. There is an inseparable bond between us that even death itself could not break." And then he tendered an offer that sounded much like the alleged pact between William James and Elwood Worcester: "If at any time either you or I should be taken out of this world I hope we can make a bargain that each of us will try to come and make contact with the other."[21]

This attempt to validate immortality may have been inspired by a desire to mitigate an old man's fear of dying, but it was also testimony to an emotional bonding to idealized parents that resisted all claims to separation. Childhood in a series of Methodist parsonages left a legacy of financial anxieties and social insecurities and few telling moments of religious certainty. The emotional assurances he remembered came through his bonding to his intense, attentive family.

Childhood Revisited: Alternate Memories

Throughout his adult life, Peale himself never separated from this image of cloudless familial harmony. He might privately allow that his father, voluble and effusive, sometimes tested the bounds of emotional restraint and that his mother, on occasion, worried and fretted too much. But all his family anecdotes in speeches, conversations, and books recall childhood with a romantic nostalgia, with even poverty remembered as a positive contribution to family bonding.

Yet many years after these early Ohio associations, having achieved success in his own right but still trying to explain—perhaps only to himself—his self-doubts about his worthiness and ability in the ministry, he interpreted his family history with a significantly different emphasis. In a written reminiscence, never published, and atypical in Peale writings for its candor, he described a home life that offered many emotional rewards: "My parents," he wrote, "were two of the happiest, most down to earth and spiritually dynamic persons I have ever known. Our home was always filled with joy, excitement, even gaiety."[22] As he now considered it from the vantage point of his own sixty years, because life within the family circle had been so comforting, it seemed to him unwarranted that its members should have been so heavily burdened by problems imposed from without: the lack of money and denominational approval for his father, the constant moving about, and the label of "preacher's kid" he had to wear himself. His resentment of these experiences, he supposed, had made a career in the ministry an uninviting prospect for him as a young man. Although in retrospect they seemed some of his "happiest memories," Peale remembered that as a boy the family lived with "stern financial stringency" and the "trying to make ends meet" policy that resulted from a salary of $450 a year, plus house, and groceries from the neighbors. As he experienced it, "the bringing in of cast off furniture for the parsonage, my Mother with all her duties struggling to maintain a home without help caused me to wonder if I wanted to organize my life around the same pattern."[23]

The petty politics of the church, which affected home life so directly, also seemed unfair and certainly unattractive. He encountered them, he wrote in language stripped bare of metaphysical romanticism, as "politics of a refined sort and as I see it now was not unnatural to man's nature even among ministers." Still, it was difficult for him to accept the psychological torment they caused his father. "But the thing that hurt me," he said,

> was to see my father caught in ecclesiastical manipulation and his spirit crushed. I learned to dread the coming of "Conference time," and listened apprehensively to discussions among the ministers as to who was "going to move" and where. There was a smug clique who had "gone to college." In those days not too many did and it seemed that there developed a fraternity of the educated. My father had graduated from medical college as he intended to be a doctor until a deep spiritual experience sent him into the ministry.

But, he continued, because his father had not attended a regular college, it was thought he

> just "wasn't quite the equal in culture" of some others. Personally, I thought he could preach rings around most of them and this discriminatory practice guised in piety disgusted my boyish and idealistic mind. I decided that if I wanted to be a politician I would go into government activity and be out and out.

Topping off all this unpleasantness was the fact that he was a "preacher's kid," at a time when that label carried heavy social implications. As a boy, he said, "I found myself resenting my status," because he was reminded of it continually by schoolmates. It came to signify

> I was set apart as a different breed forced always to carry the banner of the church, not able to be just like every other boy. It wasn't that I wanted to do

> any wrong or unworthy thing but to carry in gay and careless childhood so heavy a responsibility did not seem fair and it served to deepen an already sensitive inferiority complex.[24]

The language of this extended and unusually revealing memoir draws a sharp line between the warm and embracing haven of home and the abrasive, even hostile, environment that exists just at its borders. Written as Peale was recovering from the critical pummeling of *The Power of Positive Thinking* by academic and other scholarly writers, it is like the Janus face that observes two different scenes but from the same perspective. As a commentary on his present and immediate future, it is at once self-fulfilling prophecy and vindication of his treatment by those who deemed him not "quite the equal in culture." As an historical record of his past, it offers confirmation of a youthful impression about the tyranny of the educated elite. As his father had suffered by their stings, so had he; but unlike his father, he had by then accumulated the external resources—three university degrees and abundant financial security—to weather their attacks. Both father and son had complaints with the ministerial establishment.

A "Preacher's Kid"

As the eldest son, Norman was the personification of the Peale children to the community, and consequently he may have experienced the label of "preacher's kid" as a tighter fit than his two younger brothers. Bob, two years his junior, was comparatively less conforming than his elder path setter, although Leonard, eleven years younger than Norman, had to cope with two inherited designations, one through his father and the other through his oldest brother, whose success cast a long shadow over Leonard's lifelong frustrated efforts to find his own personal star. Norman was obviously closer to Bob, his childhood partner, than to Leonard, who figured not at all in Norman's autobiographical recollection of his youth. But it was Leonard who followed Norman into the ministry and Bob who elected to finish the medical career his father had started. To both siblings, Norman retained a responsible elder brother's role, monitoring Leonard's education and shepherding his passage through his early years in the ministry, cushioning the shocks of the role of "preacher's kid" for Bob, making it easier for him to find friends and be outgoing. Norman and Leonard were alike in sharing their mother's orderly reserve and studied public demeanor, while Bob, who came to enjoy the social whirl of the physician's life, possessed his father's gregariousness, perhaps to a fault. Bob was physically more assertive than his older brother, who, with a slight frame and timid nature, came to favor spectator sports over active competition for himself.

Norman's introduction to public school came in Cincinnati, a city then, as now, with a border state perspective that became home to many southern blacks, who, like Harriet Beecher Stowe's Eliza, hoped they were fleeing the clutches of racism. Like other northern cities, however, Cincinnati preserved a color bar. Norman's youth coincided with the flood tide of immigration, largely from central and southern Europe. Only a fraction of these immigrants settled in rural Ohio, however with a slightly larger proportion attracted to the shipping-related occupations of the Queen City of the Ohio River.[25] In Norman's first school, there were enough German-speaking children to warrant offering instruction in German in the morning and in English in the afternoon, giving him an early basis for a limited facility with the language.[26] The bilingual exposure did not last long, for the following year the family moved to Norwood, and Norman changed schools.

In his Norwood school as a fifth-grader, Norman had one of those teachers with the rare facility of leaving his mark on a young student. Known as "Professor," the instructor taught character building in a manner not uncommon among those who assumed the values of the Protestant establishment to be consistent with the nation's civil religion. His moral lessons combined Horatio Alger stories, self-suggestion, and the lessons of McGuffey Readers. After writing the word "can't" on the blackboard, the teacher would ask the class what he should do next. And, perfectly coached, the students would respond by telling him to remove the *t* from *can't*. The moral lesson—that you can change *can't* to *can* by your own effort—appealed to the fifth-grader, who stored both the story and the moral for future reference.[27] It was also in Norwood that Norman and his brother Bob formally joined the Methodist Church, becoming probationary members of their father's Grace Methodist Church on August 30, 1908. In characteristic Methodist fashion, however, since this church ritual had less meaning than the personal experience of conversion, Peale did not regard it as worth mentioning in his autobiography.[28]

From Norwood in suburban Cincinnati the Peale family returned in 1910 to small-town life in Greenville, with a population under 10,000, where they remained for three years. They were for Peale the transitional years between twelve and fifteen, the time when children typically trade the relative abandon of preadolescence for the turbulence of puberty as they struggle with questions of identity, family relationships, peer acceptance, and the opinions of the opposite sex. It was, significantly, at this point in his maturation that Peale made his life commitment to Jesus after sharing with his father his anxiety over his inferiority complex. He also reached another conclusion, this about the nature of his future life work after hearing a speech by William Jennings Bryan: Notwithstanding his fears and timidity, he had the unlikely desire to become a public speaker, maybe even a politician, himself.

He also spent a significant part of his nonschool time working at a series of part-time jobs, to see if he "really had it in him," as his father had said, to be a public person, and to earn extra money. He began as many youngsters do, peddling newspapers, delivering the morning *Cincinnati Enquirer*. Essentially a time-serving job for most kids, it was for Peale an introduction to the importance people attach to newspapers and the beginning of a lifelong fascination with the power of the printed word. His second job experience was hands-on contact with business practices, arranged by his uncle, William Fulton Peale, who marshaled the Peale social personality to promote real estate development in Des Moines, Iowa. Uncle Will, who was always a favorite with his nephew, had arranged with a bank to sell at public auction land that had been reclaimed when the owners defaulted on their mortgages. Norman's job, when he got to Des Moines, was to stake out the lots to be auctioned in towns adjacent to the city and then walk along the town streets distributing candy and an occasional $5 gold piece to advertise the sale. When he listened to his uncle talk with the crowds who gathered, he was charmed afresh by the skills of a persuasive performer, and he gained from him some common sense advice about how to address an audience, along with firsthand knowledge of the techniques of advertising and public relations.[29]

Obviously in deep conflict between the desire for public recognition and the comfort of withdrawn anonymity, Peale repeatedly placed himself in emotionally trying situations. Another part-time job put him in the demanding role of door-to-door salesman. An ad in a boys' magazine promised that for a $15 investment in a sales kit, which contained a varied collection of aluminum pots and pans and the canned sales speech to market them, it was possible for a young person to realize a modest profit. When the materials arrived, Norman discovered that the item most vital to making a sale had not come in the packet. With his initial foray into the neighboring town of Union City, Indiana, his old fears returned, and he sold nothing. Before trying again, he created with a friend a half partnership, for $7.50, and together they worked a different section of the same city. Buoyed by the companionship of his friend, he overcame his timidity and began making sales. Many years after, the man later known as "God's Salesman" acknowledged that he had gained his earliest appreciation for the art of selling from this venture with pots and pans.[30]

When his family moved from Greenville to Bellefontaine in 1913, Peale was already in high school. As a sophomore, he was once again the new kid in school, obliged to confront another set of peers, by now sufficiently skilled in the art of passing critical judgment on their classmates. Asked on one occasion to speak before the school assembly, Peale felt his knees knocking and was devastated to hear a girl in the audience comment aloud on the spectacle. The experience left wounds that even time could not heal. He refocused his interest on the more sheltered pursuit of writing.

But however much he felt himself to be an outsider to mainstream high school culture, young Peale nevertheless found friends among a small group of boys bearing such Anglo-Saxon names as Cooke, Hill, Churchill, Castle, and Newell, as well as the more distinctive Kaufmann, represented by the son of one of only two Jewish families in the town. Conspiring with a couple of these buddies, he managed the standard high school encounter with the administration, their offense a freshly painted set of numerals marking their graduation year, 1916, on the school sidewalk. Girls and dating, however, had little if any part in his high school curriculum. His sensitivity to slights, inevitable during adolescence when there are daily emotional giants to be wrestled with, endured beyond the teenage years. The rejection related to the girl's remarks about his knocking knees had its counterpart in a letter he wrote in response to an old family friend in 1951. The woman had casually remarked that as a boy Norman had been "quite a stinker," prompting him to compose a lengthy correction to dispel that "bad" characterization of him.[31] In time he came to have some understanding of the psychological roots of his sensitivity, although it had little effect on his vulnerability to perceived slights.

Peale tried his hand again at sales, taking somewhat dated models of men's suits on a horse-drawn wagon to farmers, who might consider them as prospects for a "Sunday suit," these odd jobs seeming not to interfere with his school work.[32] He finished the full high school program, which included courses in Latin and German, as well as the usual English, history, math, and science. In the summers he went with his family to the Methodist camp meeting center in Lakeside, Ohio, where the Peales were when the news broke in 1914 of the beginning of World War I. The year brought them the additional bad news that, despite effective cooperative efforts by church and civic groups, the state of Ohio defeated Prohibition by an 83,000 vote majority.

A Proper Methodist Education

The sense of national optimism about America's ability to remain neutral during the war began to slip with each new announcement of Germany's seeming intransigence. Clifford Peale's deep-seated Republicanism surely created the conviction that William Howard Taft, a son of Ohio though not a friend of Prohibition, would have made a better advocate for American interests than the intellectual Woodrow Wilson. As the country teetered on the brink of European involvement, the only comfort to be drawn from an impending crisis was, for a prohibitionist, the hope that emergency needs might force the adoption of a ban on liquor.

In 1916, the year Norman Peale graduated from Bellefontaine High School, the world still seemed reasonably safe for democracy, at least on this side of the Atlantic. With his family's encouragement he followed through on well-laid plans to attend Ohio Wesleyan University. The university was a denominational school where the children of Methodist clergy received substantial scholarship help, and where Norman, who could afford no other educational choices, had already been pledged to Phi Gamma Delta fraternity by a hometown friend. Clifford Peale drove the nervous freshman in the family car to the Phi Gam house on the campus in Delaware, Ohio, his parting message an offer of help if liquor or women ever caused problems—a generous offer, but unlikely prospects on a dry campus that forbade even dancing.

Ten years before Peale entered Ohio Wesleyan, Ernest Fremont Tittle had graduated from the institution. Tittle went on to become a leading figure in the liberal movement, in theology as in social action, not only in the Methodist Church but in Protestantism generally, voting on occasion as a Christian Socialist. Years later, when writing to a friend about the school that inspired his decision to enter the ministry, Tittle said,

> As undergraduates you and I were indeed exposed to a certain kind of idealism, yet as I now view it it was a little sticky with sentimentalism. At any rate, it did not come to grips with the political or economic situation at home or abroad. Would it be going too far to say that it was almost wholly devoid of social insight? It was largely "missionary," that is, a compound of unselfish devotion and spiritual pride.[33]

From his politically advanced position as a social activist, Tittle could pick out the limitations in his undergraduate education, yet his comments with only slight modifications applied equally to most denominational colleges and universities of the period. They were seen by their benefactors as training grounds for the faithful: for ministers, stalwarts in the churches, and defenders of the Protestant establishment. Ohio Wesleyan had been accomplishing just that since its founding in 1844, and had it deviated from its mission, a watchful board of trustees would have called it to account.

Evangelical and sectarian, the university followed the accepted Methodist code of conduct by banning the customary trilogy of smoking, drinking, and dancing and by requiring attendance at weekday chapel and Sunday church services. Virtually all the students at the university were Protestants, and the majority of these were Methodists, at the time the largest denomination in the nation and in the state.[34] Revivals, prayer meetings, and hymn sings supplemented the curriculum and also served as social occasions. In 1916 when Peale arrived on campus, the Reverend Dr. John Washington Hoffman was beginning his first year as the school's president—a coincidence the two men would remark on often over the course of their lasting friendship—and his tenure was expected to produce an updated curriculum, staffed by new faculty recruited from developing fields within the social sciences. The school had a history of sending an impressive number of its students to some of the best graduate schools in the country—a mild rebuke to Tittle's impression of a "missionary" atmosphere—and Hoffman's agenda was supposed to hone its competitive edge.[35]

If the student body of Ohio Wesleyan had the homogeneity of small-town Ohio, it was a more congenial environment for Norman Peale than public high school because of its patent identification with Methodism. Life in the fraternity house produced its share of boyish hijinks, serving up daring, though harmless, initiation rites, which once sent Peale on a midnight trek through the countryside. A surrogate family, the fraternity promoted camaraderie and created the basis for friendships that Peale, with his sense of familial loyalty, cultivated throughout his life. It fostered the kind of intimate atmosphere in which he could thrive, and in his last year as an undergraduate, he was president of the local chapter. It may even have inspired enough social confidence for him to develop a relationship with a woman student, although it is a subject on which he maintained a public and private silence.[36]

The modified curriculum brought in the expected new offerings in the social sciences while keeping in place the traditional class format of lecture and student recitation. Whatever feelings of social liberation Peale felt in the fraternity house failed to sustain him in the classroom, where he continued to be haunted by his old fears when called on to recite. On those painful occasions, his shyness would overwhelm him, and his face flushed with embarrassment, he would stumble over answers he knew. His economics instructor, a Professor Arneson, informed by either years of experience or possibly the new psychological literature of narcissism, detained him after one of his classes to talk about the problem. Arneson accurately and knowledgeably observed that an inferiority complex was actually rooted in egotism and self-centeredness. Aware of his student's background, Arneson sent him on his way with two potent suggestions: to ask Jesus to help him deal with his problem and to check the writings of William James in the library.

The confrontation was devastating for Peale, but the outcome was as it had been earlier in a similar, if gentler, exchange with his father. After moments of empty rage, he later recalled, he went to the chapel steps, where he prayed that Jesus would "change a poor soul like me into a normal person" and immediately, caught in the soothing embrace of the "Presence" of Jesus, presumably found confirmation that his prayer had been answered.[37] Then, shortly thereafter he was guided by his professor of English literature, William E. Smyser, to works by Emerson and the *Meditations of Marcus Aurelius* for a sympathetic unfolding of the power of the individual mind.[38] Smyser's personal fondness for the Romantic tradition in literature, particularly the poetry of Wordsworth, was also transmitted to Peale, and his classroom reading of the "Intimations of Immortality" evidently produced for student Peale the kind of epiphanic experience all teachers dream of.[39]

Peale's discovery of James and Emerson, and to a lesser extent Marcus Aurelius, acquired in the atmosphere of romantic idealism that seemed to flourish on the Methodist campus, eventually became part of his mental equipment and then a lifetime fascination. He would soon encounter the Emerson of Transcendentalism again in seminary as a shaping force in liberal theology, and James's views on the power of the unconscious mind would provide the basic text for many of his courses in social theory.[40] This initial encounter with the renegade Emerson and the iconoclast James, had it been only a passing recognition of a college student with romanticized kindred spirits, would have failed the test of time when Clifford Peale's version of the "real world" intruded. They remained enduring intellectual companions because they counted for something in Peale's own life.

The Jamesian-Emersonian metaphysical spirituality that penetrated the Ohio Wesleyan atmosphere, evidenced in the references of Peale's Professors Arneson and Smyser, was a byproduct of a more general faculty appreciation for the mystical philosophy of Personalism as represented by Methodist theologian Borden Parker Bowne. The material recommended by his instructors provided a form of intellectual authority for an inherited Methodist sense of the mystical phenomenon of the Presence, a view that his parents shared. Romatic metaphysical liberation was also "in the air," for although the Emmanuel Movement had passed its prime and the prewar ethereal "liberation" era to which historian Henry May referred had fallen victim to national preparedness, the waning of sympathies was neither uniform nor precipitous. It was clearly still in evidence in some departments on the Ohio Wesleyan campus.

It also seems evident that this early encounter with James and Emerson was auspicious for Peale; he would often return to their views for validation as his ministry unfolded. James and Emerson became for Peale more than disembodied intellectual guides. Emerson, the self-proclaimed "practical idealist," provided the language for dealing with some of the discontinuities and polarities Peale was beginning to recognize in his own life—for the struggle, for example, between service and satisfaction, between an orthodox notion of self-denial and a modern understanding of reward. Peale also found Emerson's sociological perspective as compelling as his theology: The Concord sage could simultaneously endorse great men and indict intellectual arrogance; he could celebrate nationalism while extolling the mysterious East; he could welcome commerce but revere pastoral Nature; he could herald the arrival of modernity but fear the tyranny of science. There were similar kinds of paradoxes in Peale's own life—contradictions about authority, personal worth, sacrifice, security, and intellectual competence—and Emerson provided a rationale for understanding these conflicts. Peale had no dramatic conversion to Emersonian views, any more than to Jamesian ones, but over the years they became increasingly important references, sources in whom he placed significant confidence as authorities for his version of Practical Christianity.

In the case of James, the influence was psychological and pragmatic. Peale did not begin his ministerial career preaching sermons of practical optimism:

that came later, near the close of the roaring twenties. But James's optimistic "healthy-mindedness" was an easy match with Clifford Peale's belief in a benevolent future, a view premised more on fortuitous Nature than on material circumstances. James's understanding of behavior and the role habit played had obvious personal relevance for the younger Peale in coping with his nagging inferiority complex and in charting a way around it. If habits became ingrained through their repetition, then to continue to hide and try to make oneself invisible to deal with shyness would only perpetuate the problem. By acting self-confidently, in time the pose would become the new habit, and the system would be purged of the old demon. And James described the abundant inner resource, the power of unconscious mind or the positive subliminal self, available to manage the transformation. In time, the Jamesian ideas found so comfortable a niche in Peale's thinking that it was a well-intended observation when his wife of many years remarked that he cited James more than Jesus in his sermons.

Just how seriously the Emerson-James incarnation affected Peale's college career is not clear; like good aged cheese, it seems to have been an acquired appreciation, its value enhanced over time. Peale was an English literature major, his reading dictated by professorial preferences for the English Romantic writers. But his pragmatic interests drew him to the social sciences, and he took as many courses in that area as he did in his major. Peale was predictably a hard-working student, whose grades were good but not outstanding, his recitation fears obviously taking a toll in some areas. Some of his worst grades were in the field called oratory, where he received a surprising A in extempore speaking, but C's and D's in principles of elocution and literary analysis and interpretation.[41]

Planning for the Future: A Career or a Calling?

Mixed in among Peale's youthful dreams and continuing for a number of years thereafter was a life in active politics. For a time the dream took on life when he agreed to organize the campus campaign for Republican presidential aspirant Leonard Wood, a consuming volunteer job that required his taking a trip to the Ohio State House. When he got to stand within the state house rotunda, with its monuments to Grant, Sherman, Sheridan, McKinley, and similar notables, he noticed a strikingly bare place, in which he found it easy to imagine himself. When he thought about that visit in retrospect, he said,

> My fond ambition was that some day this niche would be occupied by a statue of Norman Vincent Peale, another of Ohio's Jewels, an immortal political figure of the Buckeye State. I didn't exactly aim for the Presidency, but I was not unaware that in those days Ohio was known as the "Mother of Presidents."[42]

Peale had been raised on political commitments held almost as dearly as religious convictions, and he savored the intensity and relished the excitement of politics. In time however, he discovered that he was not cut out for the rough-and-tumble confrontation of partisan politics. He worked diligently on behalf of presidential-hopeful Wood, for whom his loyalty ran deep, as it would continue to do whenever he took on the advocacy of a particular candidate.

Most of his undergraduate extracurricular activities, except for the war years, were more parochial than the arena of party politics, extensions of his dual ambition to pursue a career in writing and/or to succeed as a public speaker. In his sophomore year, college life grew suddenly and uncharacteristically serious. In April 1917, President Woodrow Wilson announced America's entry into the war, and Peale enlisted in the Students Army Training Corps, or ROTC. The students were uniformed and expected to be combat ready, but the war's end terminated their services before they had to demonstrate their skills, and the more usual pattern of campus sociability returned. Peale was honorably discharged from the Students Corps in December 1918, possessed of a record that would later give him entry to membership in patriotic veterans' groups.

During the summer university recess, he returned to his family's new home, now Findlay, Ohio, and in the interval before the start of his senior year polished up his writing credentials to apply for a reporter's job with *The Morning Republican*, the local newspaper. His first assignment as a cub reporter was the obituary page, and as his talent got noticed, he was promoted to police cases and other city news. He enjoyed the recognition, and began thinking more seriously about writing as an acceptable career. When a classmate suggested that they make the leap into entrepreneurship and together own and run a small-town newspaper, he caught a vision of an idyllic future; a self-governing writing job that could simultaneously provide him with a comfortable nest egg. He still rankled at the memory of his father's patronizing treatment by the local banker, who once asked smartly as he handed the preacher his small stipend, "Do you think your sermon justifies this?"[43] And he hoped that he would not be forever obliged to drive the kind of least-expensive, secondhand cars that his parents had to settle for, but might one day be in a position to buy a new Cadillac, as one of his father's church members had recently done. But for two college students with no financial resources, the prospect of becoming owners of an independent newspaper was beyond the realm of possibility, and the entrepreneurial dream died.

The hope for a writing career, however, remained very much alive, and so after receiving his B.A. degree from Ohio Wesleyan on June 16, 1920, Peale packed his belongings and headed for a reporter's position with the Detroit *Journal*. It was far from his last contact with the Delaware campus; he would return many times in the future for university functions, and in 1930, to collect an honorary doctor of divinity degree from his alma mater.

The young graduate had a brief, satisfying tenure with the *Journal*, a fifteen-month stint that he said left him with a lasting attraction to "printer's ink." During his time in Detroit he cemented a relationship with one of the editors, Grove Patterson, who gave the novice reporter some practical lessons in journalism, including a maxim that was readily incorporated into the treasury of Peale lore. It was Patterson's conclusion that the period was "the greatest literary device known to man," with short, punchy sentences the most effective form of written communication.[44] The older man's advice was to use a simple, uncomplicated language readily understood by the most ordinary reader, such as a "ditch digger." Peale's fondness for Patterson and his ready acceptance of the editor's populist, unstuffy approach to the craft became new reasons for rationalizing writing as a vehicle for service, but he was still ambivalent about it as a life's work.

Then the irresistible "call" to the ministry finally came, forced to the surface in a reassuring, emotional setting not unlike the circumstances surrounding his arrival into the world. The prospect had never really been out of his mind, requiring only the appropriate stimuli to end his indecision. After a year in Detroit, he decided to join his parents at the annual Methodist Church Conference, held that September in the familiar surroundings of the Delaware campus. Back in that supportive environment, with his parents, the sermons, and the hymn singing, the old prompting to enter the ministry returned. Feeling especially moved by one of the sessions, he went outside to walk and to think about his choices; he enjoyed writing, was fascinated by politics, but there was that unrelenting tug. There were also the expectations of his parents, unspoken, yet understood. On Sunday night, the Conference concluded, Peale returned with his parents to where they were staying, in his boyhood town of Bellefontaine. Still uncertain, he walked past some of his old haunts until late at night, when finally, as he remembered it, the conviction came, illuminated by the Presence of Jesus.

The decision settled, he decided to act on it, quickly and decisively, by sending a telegram asking Boston University whether it was still possible for him to apply, since it was already mid-September. With the main telegraph office closed, he realized the only possibility for getting his message out was the dispatcher's shack, a tall wooden structure that also served as a water tower, built close to the railroad's edge. He climbed the high wooden ladder to the small office, and the obliging attendant sent his request out on the wire.[45] The next day, a telegraphed acceptance arrived from Boston, the first and only school to which he applied.

Still, Peale hedged his bets. His primary enrollment was in the master of arts program in literature at Boston University, with room in his schedule to take courses in the university's School of Theology, then a notable hotbed of liberal opinion. Before his first academic year had ended, he switched his enrollment to the seminary, while continuing in the master's program. His commitment to the ministry was final.

It is not necessary to unpack a nature-nurture argument to appreciate the extent to which his decision had been influenced by his family background and unique life circumstances. The authenticity of his call is in no way diminished because historical circumstances had positioned him to receive it. Indeed, it could be argued that there were compelling psychological reasons for rejecting the ministry and pursuing his classmate's suggestion to go into business. Retrospectively, however, his decision appears virtually inevitable. Parental expectations, a powerful paternal role model, the seamless quality of his relationship to his family and from them to the church, all were factors working against any other vocational choice. Furthermore, while being minimally acquainted through his job experiences with the cultural separation of the sacred and the secular, Peale was more than used to a role definition imposed by the church; to be part of a nonurban Ohio Methodist preacher's family was to have one's life defined with unusual specificity. The ministry would give him a protected arena in which to cultivate his public speaking desire, and as for writing, there were always sermons, if nothing else.

The specter of clerical poverty, he admitted, clouded and postponed his decision, partly banished, perhaps, by a subconscious awareness of the prospects for greater psychological rewards. It is not uncommon among those who choose a religious vocation to see in service to the metaphorical Mother Church a reflection of their personal family history, to view the church as either an extension of their own idealized family or as a substitute for a difficult one. For Peale the fusion between family and church seemed so secure as to be impossible to separate. It might have been experienced, too, as a nurturant place in which to work out his preoccupation with his inferiority complex, with its narcissistic, or as his professor termed it, self-centered, quality.

Included in his luggage as he left for seminary was a secure Methodist heritage, which mixed with a relish for the kind of partisan politics associated with the Protestant establishment: not notable qualifications for a future metaphysical healer. There was, to be sure, a strong mystical, pietistic Methodist tendency in him and in his family. He had been reared in an atmosphere offering the unusual combination of pietism and pragmatism, and it required education and experience for him to connect his Emersonian-Jamesian inspiration with his dislike of "big shotism" and his own psychological struggles. Only then did he discover in the resources of the unconscious a home remedy that "worked" for him and would find increasing favor in popular culture.

Notes

1. Obituary, Sept. 21, 1955; Norman Vincent Peale Manuscript Collection, Bird Library, Syracuse University. Hereafter cited as NVP Ms. Coll.

2. William R. Hutchinson, *The Modernist Impulse in American Protestantism* (New York: Oxford University Press, 1982), p. 43.

3. Quoted in Paul A. Carter, *The Spiritual Crisis of the Gilded Age* (DeKalb: Northern Illinois University Press, 1971), p. 8.

4. Sydney Ahlstrom, *A Religious History of the American People* (New Haven; CT: Yale University Press, 1972), p. 740; William G. McLoughlin, *Modern Revivalism: Charles Grandison Finney to Billy Graham* (New York: Ronald Press, 1959), pp. 219–20, 221–81.

5. See George M. Marsden, *Religion and American Culture* (New York: Harcourt Brace Jovanovich, 1990), especially pp. 53–110.

6. Quoted in McLoughlin, *Modern Revivalism*, p. 259.

7. Ibid., p. 253.

8. Ernest Cherrington, *The Evolution of Prohibition in the United States of America* (Montclair; NJ: Patterson Smith, 1920). Cherrington was a participant in many of the events he recorded. See also Sean Dennis Cashman, *Prohibition* (New York: Free Press, 1981).

9. Norman Vincent Peale, *The True Joy of Positive Living* (New York: Morrow, 1984), p. 18. The book is subtitled *An Autobiography.* Like all of Peale's writings, it is essentially anecdotal and is seemingly drawn partly from notes and partly from memory. As a result, there are discrepancies between dates Peale cites and those given by his authorized biographer, Arthur Gordon, in his *One Man's Way.* Correspondence in the NVP Manuscript Collection also reveals some differences, relating largely to dates, with the accounts Peale gives. When asked about differences between his dates and Gordon's Peale was convinced his own were accurate. The autobiography contains a lengthy report of his early family life, which he describes as largely idyllic.

10. Hardly an unusual childish reaction, though slightly out of keeping with the image of the close and loving family Peale liked to convey. The account was, surprisingly, included as part of a family display—pictures and commentary—in the new library of the Center for Positive Thinking in Pawling, New York, in June 1989.

11. Peale, *True Joy*, p. 25.

12. The anecdote, as written and spoken, is the Peale stylistic form of expression par excellence. His private conversation is as anecdotal as his public speaking and writing. Part of the reason for his preference is that he is good at it, he is a gifted storyteller, and part of it is probably for reasons of self-protection, as a screen against uncomfortable closeness. It is also true that the narrative style frequently places limits on the ability to be analytical. Peale's method of illuminating an anecdote is with another anecdote. Because his conversational and autobiographical memoirs form the primary, and essentially only, source for his early life history—there being no documentary evidence for his early years in the NVP Manuscript Collection—the record inevitably abounds with Peale anecdotes.

13. Peale, *True Joy*, p. 31. The incidents here are taken from the autobiography, but they are staples in his sermons, speeches, and books. They also recur in conversa-

tion, as during an interview on December 29, 1980, when he repeated the story of Dave, the town drunk.

14. Interview with Norman Vincent Peale, Pawling, New York, July 7, 1989. Immediately after telling this story, Peale told another, about a seventy-year-old, white-haired woman in his congregation at Marble Collegiate Church who once came to him for counseling. She confessed that as a young woman she had been asked to have sex with a man whom she liked. Although she refused, she had been plagued since then by guilt, because although she had said no, she really wanted to say yes. Peale asked if she believed that as a minister he could act for God and forgive her; she said yes. Peale became very tearful, and choked up as he continued the story. They went into the church sanctuary, and as she knelt down, he put his hand on her "snowy white head" and said she was now as pure as when she was born. Peale was ninety-one years old when he told these stories, and there was a great deal of nostalgia involved. It is not difficult to imagine that the tears for the woman reborn were subconsciously mixed with memories of his own youth.

15. Psychiatrists claim that when the oedipal separation is elusive, a boy may find in his father his own ego's ideal, a surrogate for the state of harmonious union with the maternal figure. One student of the subject has described the ego ideal as the Mona Lisa counterpart to the superego's Simon Legree, an alluring, often romatically deceptive, invitation to live harmoniously with the ideal, now and in the future. I have found the concept of the ego ideal a helpful psychological tool in trying to unravel some of the paradoxes in Peale's makeup and his message. See Janine Casseguet-Smirgel, *The Ego Ideal* (New York: Norton, 1985), pp. 26, 27; John M. Murray, "Narcissism and the Ego Ideal," *Journal of the American Psychoanalytic Association*, 12 (1964):477–511.

16. Peale, *True Joy*, p. 14.

17. Letter of C. C. Peale to NVP, Apr. 1927; NVP Ms. Coll. In the letter to his son, Clifford Peale described his wife's restlessness: "She is weary and tired of the pastorate. I would to God some help might be found so that I could give her relief." In letters to his son during the 1930s, Clifford often wrote of Anna's declining health and sagging spirits.

18. Letter from Mrs. S. R. Garrison to NVP, May 7, 1952; NVP Ms. Coll.

19. Letter of NVP to Mrs. H. C. Weaver, Mar. 26, 1951, NVP Ms. Coll.

20. Peale, *True Joy*, p. 21.

21. Letter of NVP to C. C. Peale, May 29, 1951; NVP Ms. Coll.

22. Norman Vincent Peale, handwritten recollection, n.p., n.d., though probably written in 1958; Box 300, NVP Ms. Coll. This is a lengthy document, with corrections made by Peale. There is no indication of the purpose for which it was written. It reveals greater self-examination than anything else in his voluminous manuscript collection and may possibly have been intended for presentation before some member(s) of his church's therapeutic clinic, the Institutes for Religion and Health. In 1958, Peale remained in the midst of a critical storm regarding the publication of *The Power of Positive Thinking*, and this may also have contributed to his examining his original decision for the ministry.

23. Ibid.

24. Ibid.

25. Robert Moats Miller, *How Shall They Hear without a Preacher: The Life of*

Ernest Fremont Tittle (Chapel Hill: University of North Carolina Press, 1971), p. 7. In developing his biography of Methodist minister Tittle, who was born in central Ohio fourteen years before Peale, Miller presents a useful and engaging picture of the social and economic life of the state in the late nineteenth and early twentieth centuries, the years of Peale's youth.

26. Letter of NVP to S. Olson, May 26, 1956; NVP Ms. Coll.

27. Peale, *True Joy*, p. 31.

28. Letter of F. L. Cunningham to NVP, March 1, 1950; and NVP to F. L. Cunningham, Mar. 22, 1950; NVP Ms. Coll. The letter to Peale indicates that the boys were received into membership "on probation," which preceded full membership. Peale's letter in response says, "I had forgotten the date on which I became a full member of the church and was very happy to have you tell me that it was on August 30, 1908." There is no evidence that Peale had one single, soul-searing conversion experience, but rather successive contacts with what he termed "the Presence." The episode with his father in the countryside, when he vowed to give his life to Jesus, appears as one of several similar decisive events in his life. Since the commitment ritual with his father occurred after he became a probationary member, it is possible that was the "saving" experience which qualified him for full church membership.

29. Peale, *True Joy*, pp. 40, 41.

30. Ibid., pp. 44–46.

31. NVP to R. R. Thomas, Apr. 11, 1951; NVP Ms. Coll.

32. Peale, *True Joy*, pp. 51–53.

33. Letter from Tittle to Harry Gorrell, Nov. 13, 1935; quoted in Miller, *How Shall They Hear without a Preacher*, pp. 21, 22.

34. Miller, *How Shall They Hear without a Preacher*, pp. 19, 20. Miller notes that Methodism had twice the membership of the next largest denomination, the Presbyterian.

35. Although Peale attended Ohio Wesleyan a decade after Tittle, they had some of the same instructors. Peale's residence in a fraternity house made an important difference in the social experiences of the two men.

It should be noted that in 1916, the year Peale arrived at the university, Tittle lost to Hoffman by one vote the opportunity to be the school's next president, which may help explain Tittle's wry comments about his undergraduate education. In *How Shall They Hear without a Preacher*, Miller offers an informed discussion of Ohio Wesleyan during these years; pp. 50–51.

36. Letter of Leonard Peale to NVP, Nov. 2, 1932; NVP Ms. Coll. After referring to his own serious interest in his girlfriend, Leonard said, "I also remember hearing about your affair in college, and I am trying to avoid any such results."

37. This story, with many variations, is a staple in Peale's books and sermons. See *True Joy*, p. 57.

38. Miller points out that Smyser had earlier been responsible for persuading Tittle to consider the ministry. See *How Shall They Hear without a Preacher*, p. 23; and Peale, *True Joy*, p. 57.

39. Letter of NVP to Charles L. Wallis, Apr. 20, 1949; NVP Ms. Coll.

40. The Emerson-James-Peale connection is also illustrated in Donald Meyer's critical discussion of Peale: *The Positive Thinkers* (New York: Doubleday, 1965).

41. NVP college transcript, Ohio Wesleyan University, dated Sept. 27, 1923; NVP Ms. Coll.

42. *The Nation's Business*, August 1970.

43. Interview with NVP, Pawling, New York, February 24, 1983.

44. Peale, *True Joy*, p. 69. This anecdote is another of the staples in Peale's public offerings, written as well as spoken.

45. Telephone conversation with NVP, August 19, 1987. In his biography of Peale, Arthur Gordon locates this struggle with the call in Bellefontaine, while Peale had situated it in Delaware. Peale clarified the seeming discrepancy by pointing out that the decision consumed several days, beginning in Delaware and continuing in Bellefontaine. In his conversation, he referred again to climbing the stairs to the water tower, the deliberateness of it all somehow sealing his commitment.

CHAPTER 2

Learning the Lessons of Liberalism: 1921–1932

Life is not dialectics. We, I think, in these times, have had lessons enough of the futility of criticism. Our young people have thought and written much on labor and reform, and for all that they have written, neither the world nor themselves have got on a step. Intellectual tasting of life will not supersede muscular activity. If a man should consider the nicety of the passage of a piece of bread down his throat, he would starve.

Emerson
Essay on Experience
1844

I am come that they might have life, and that they might have it more abundantly.

The Gospel of John 10:10

As Anna and Clifford Peale stood at the Toledo train station waving goodbye to their Boston-bound eldest son on that September day in 1921, they surely had few misgivings over his career choice. There were personal costs attached to heeding the call to the religious life, but they had themselves snagged the lion of sacrifice and knew it to be harmless. For them, the social exchange—the warmth and regard of a local community, status, the personal sense of service—was worth whatever financial compromises had to be made. They had encouraged Norman's decision for the ministry, and they were satisfied that the religious heritage he took with him to the hotbed of liberal theology was as much their own as the box lunch Anna had carefully packed for the trip East.

The future eventually confirmed their confidence. Norman Peale emerged from the seminary an evangelical liberal, politically more conservative than Elwood Worcester had been, but with a similar commitment to a ministry of personal religion. His seminary experience turned out to be a rough initiation to the profession, its assertive liberalism as alienating to him as its competitive environment was unexpected. So unpleasant was the time he spent there that he returned only once after graduation, and it was a predictably trying occasion. It seems likely that his enduring resentment of academics generally, but seminary professors in particular, owes something to these years.

The times favored Peale's hopes for the future, at least superficially. He early harnessed his ministry to a cultural trajectory, and as the prosperity decade of the twenties seemed daily to fulfill its promise, he soared with the social ascendancy. The opportunities turned out to be ephemeral, but he made the most of them. The decade opened with the defeat of the League of Nations in 1920 and ended with the stock market crash of 1929, but the years between were defined by more colorful and exciting events. Mainstream America yearned for a life less restricted by old conventions, more open to new possibilities in sexual life, automobile travel, consumer goods, education, and the youthful pleasures of the jazz age. Close to the pulse of the culture, Peale understood these desires and found ways to give them expression in the programs of his local church. By the time the prosperity bubble burst, leaving activist liberal churches with budgets they could not meet, Peale had reorganized his theological priorities to accommodate the mental healing emphasis of the metaphysical tradition. It helped his ministry to flourish during the dark days of the thirties, as other mainstream churches struggled to stay open.

Ironically, Peale's unpleasant experience in seminary worked in his favor during this period of uncertainty for collective religious activities because it exacerbated an almost instinctual suspicion of forms of authority, inside or outside the church. Basically a loner, he fit the stereotype of the person comfortable with a crowd of 4,000, but ill at ease with four. When he worked at cultivating social interaction, Peale could be very good at it, but social intimacy was not easy for him. He and organized Protestantism and agencies of ecumenism seemed to develop a tacit agreement about keeping mutual distance from each other, thus giving him the freedom to work at the margins of the mainstream tradition.

Protestantism generally followed a similar boom-and-bust pattern as the rest of the culture during the decade, which Peale viewed as a depressing commentary on its basic lack of vitality. Historians of religion usually point to the same auspicious beginnings: the adoption of the Eighteenth Amendment to establish Prohibition and the creation of the Interchurch World Movement in 1918. But by the end of the decade, there was clear evidence that Prohibition would have but a short future, and the Interchurch World Movement was already dead. Prohibition was Protestantism's last, most successful united effort, in many ways a tribal fight to demonstrate the enduring quality of Protestant hegemony. Because so much was invested in the campaign in terms of group identity, the effect of the threat and then the reality of repeal was profound: It fed feelings of disillusionment and frustration with the church, at a time when boredom and disaffection were coming to characterize more members' attitudes toward their local parishes.[1]

Although there was abundant evidence that the old religious progressivism continued to prevail, by the end of the decade ominous signs were everywhere of the fragility of faith among Protestants. Church attendance was in decline, mirroring a similar falling away by those who had previously been attracted by the Billy Sunday revivals, and the disaffection of intellectuals continued. There was a dramatic drop in giving for missionary purposes and fewer candidates for the work of foreign missions. The apparent slide in commitment had connections to the vibrancy and intensity of an alternate cultural message, which was presented to the public in such products as Bruce Barton's 1925 best seller, *The Man Nobody Knows*, which pictured Jesus as a successful business manager, and which, in addition, was more fully revealed in "the shift in American life that was displacing evangelical Protestantism as the primary definer of cultural values and behavior patterns in the nation."[2] The question of how relevant religion could or should be was a vexing one to those in mainstream churches who thought seriously about such matters.

Before the close of the decade, through ways he never fully revealed, Peale developed a clear appreciation of the need for cultural relevance and the alternate tradition of metaphysical spirituality. The forms of what was known as religious science, which first attracted him, were redefined in the twenties through the popular offerings of such inspirational writers as Ralph Waldo Trine, Emmet Fox, Glenn Clark, and the French pharmacist Emile Coué. Martin Marty has suggested that as early as the tribal twenties, it was possible to discern the configuration of not just three major faith groups—Protestant, Roman Catholic, and Jewish—but four, which included a Religion of Democracy, or a religion-in-general.[3] In Marty's view, this fourth force derived from the alternate metaphysical tradition as revealed in Christian Science and New Thought and its popular reinterpreters such as Trine and Fox.

"There Is Too Much Philosophy Here"

Peale handled the external adjustment to life in Boston with ease. He moved into rooms, first on Mt. Vernon Street and then on Louisburg Square, and found a part-time job running a dumbwaiter at the Beacon Street YWCA. His recreation was usually the standard seminarian excursion of going to hear the sermons of one of the local clergy or some distinguished visiting preacher. He made a particular effort to hear the inspirational healer Emile Coué—"Every day in every way I'm getting better and better"—and the message, he reported, "intrigued" him. Once a week during his last two years in theological school he arranged to have dinner with his brother Bob and Bob's friends, all students at Harvard Medical School, his primary source of socialization.

His adjustment to the intellectual atmosphere that drifted over the Charles River was, however, more problematic. At the end of his three-year program, as he was about to exit the seminary, he reflected on his experience in a questionnaire distributed by the school, and his judgments were not particularly kind even then. That he was a senior just about to face life in the "real" world undoubtedly colored some of his remarks, particularly his feeling that the curriculum was largely irrelevant to the practical demands of a minister. In the biography of the Reverend Ernest Fremont Tittle, a man fifteen years Peale's senior, Tittle is reported to have had earlier expressed similarly negative feelings about his education at Drew Seminary, another Methodist institution.[4] Peale had followed the standard seminary program; courses in the Old and the New Testament, church history and Methodism, systematic theology, religious education and psychology, practical theology, social service, and missions and had earned a B+ average. Peale seemed to do best in courses in the social sciences, as he had in college, while continuing to struggle with public speaking.[5]

But the critical comments he made on the 1924 seminary questionnaire also had a prophetic quality. He expected to have an urban ministry, he wrote. He was disappointed, however, that the seminary had not prepared him better for his life's work, that it had not helped him to figure out how to harmonize the scripture with "the knowledge of the present" and "to enable me to go out and preach—to strengthen and amplify the message which I feel I have." The problem with the seminary's offerings, he noted, was that

> There is too much philosophy here—we need a good grounding in philosophy to be sure but oftentimes the main interest appears to be directed toward exalting the ideas of Borden P. Bowne et al. I feel that we need a greater stress on the practical work of the ministry. The world needs more preachers rather than philosophers important as they are.[6]

In response to a specific question about how the school had failed to realize his expectations, he said, "The school has certainly grounded me well—it has taught me *what* to say but not so well *how* to say it. I should have liked more Homiletics."[7] The case for more homiletics seems an overstatement, given the strongly negative comments he eventually made about the preaching courses he had taken. Much later in life, he said that a course in psychology of religion, which he had taken with Professor David Vaughn, had been the most useful to his life's work.

His basic quarrel was with what he regarded as the seminary's dual emphases: the focus on the social gospel imperatives of Walter Rauschenbusch and other theological liberals, and the seeming preoccupation with the metaphysical Personalism of Borden Parker Bowne. Peale arrived in Boston with a negative predisposition toward the social gospel, and nothing over the next three years occasioned a change of heart, except perhaps to deepen his original conviction. With Bowne, he had no equivalent history.

Like Peale a Methodist, Bowne had taught at Boston University and was brought to trial for heresy by his conference in 1904 before being acquitted. His name became synonymous with Personalism, sometimes referred to as Boston Personalism, and for the first three decades of this century his ideas permeated the Boston University School of Theology and similar institutions. It was said that "Bowne's personal idealism was almost the official philosophy of American Methodism for nearly a half-century."[8]

It is possible to conceive of Bowne's philosophy of Personalism as the intellectual mystical relative of that emphasis on inner spiritual intensity associated with the nineteenth-century Methodist Holiness movement. While Bowne challenged the Social Darwinism of Herbert Spencer and tried to build a bridge between his own ideas and the social gospel, he was more concerned with the inner life, with abstract, metaphysical notions of mind than with issues of social ethics. He regarded mind as the fundamental reality, containing "principles immanent" within it that reflected nature, experience, and spirituality.[9] Personality was the manifestation of this concept of mind, and for Bowne, it was the Personality of Jesus, his essence, that was most significant.

At Boston, Peale studied with Albert C. Knudson, "personalism's chief systematic theologian," who taught a seminar on Schleiermacher and Ritschl to demonstrate his general sympathy with these German theological liberals, but more especially to describe the difference he saw between Boston Personalism and Ritschl.[10] Another influential Personalist at Boston, Edgar S. Brightman, was also one of Peale's instructors, who, though he studied Ritschl in Germany and wrote his dissertation on him, rejected Ritschl's notion that religious faith was independent of metaphysics. Brightman evolved a qualified monism, with a high regard for mysticism, offering a course on the classics of mysticism.[11]

Peale's course of study at seminary was therefore a mixture of theology, philosophy, and social science, of the mysticism of Personalism and the activism and ethics of the social gospel. His parting comments to the field work office on the nature of the curriculum are therefore revealing for their singular focus on overexposure to philosophy. Uncertain about how to receive and interpret Personalism, finding it on the one hand stodgy and overly analytical and on the other uncomfortably close to concepts he identified as his own, it became another means for nurturing a metaphysical subjectivism that had been planted in his religious outlook in his earlier days.

Peale saved many of his notes and papers from his seminary years. The only evidence of his attempt at serious scholarship, they provide important commentary on how he received the philosophical fare of his day and how he was subsequently sustained by it. It became almost a cliché with him in later life to say that theology when he studied it was in a "confused" state and a "mess," although he confessed that for someone like himself without a "speculative bent" it was intellectually too strenuous. His papers therefore did not grapple with the hard stuff of Bowne-Knudson-Brightman, although their perspective is represented in the work. In a general way, the papers portrayed a Bowneian appreciation for the personality of Jesus and the reality of mystical experience, along with a willingness to accept the subjectivist position as against a materialist or pragmatic modernist one. One of his most forceful papers was a critique of the ideas of John Dewey.

Peale's newspaper background had taught him to write clear and lucid prose, a skill that showed to best advantage when he was writing about historical or social subjects. For one course he composed a predictably sympathetic discussion of John Wesley and George Whitefield, but then also developed a surprisingly supportive analysis of Rauschenbusch, although he pointed out that he thought the noted liberal's policies were politically "unrealistic" and mistaken in their judgments about the defects of capitalism. As one who had endorsed the League of Nations, he took issue with the prominent Marxist Harry F. Ward's observation in his *New Social Order* that the League was a "holding company for capitalists," although in an interesting switch he thought the rest of Ward's proposals offered a "fair" and "practical" plan for "a new social order." The paper that best represented Peale's theological priorities as a seminary student was "Had Jesus a Social Gospel?" a question he answered affirmatively. He described Jesus as a "social settlement worker" whose ministry to individuals in the mass might best be described as "Social Individualism," which was a fairly accurate term for his own evangelical liberalism.

The most thoughtful paper he wrote in seminary, developed with authority and strength, dealt with the textile strike then in progress in Rhode Island and affecting the community where Peale was serving at the time as a student pastor. After presenting the issues as both labor and management saw them, he situated himself on the side of labor. Given his later close ties with business executives and management, this early claim that labor is less able to withstand the consequences of a compromised settlement revealed his rapport with the views of his local congregation. The best strategy for mediating such disputes, he contended, was to have management state its case truthfully and sensitively to labor, thereby enabling labor leaders to have a more realistic picture of what was at stake. Buried quite deeply in the discussion is an implicit desire to "harmonize" these traditional foes, but student Peale concluded that if sacrifices had to be made, management should make them.[12]

When he left seminary, he described himself as a liberal, although inoculated against the social gospel bias of his professors. His opposition was hardly rabid, however, and in any conflict with fundamentalists his spontaneous reaction was to side with the modernists. Given a choice, he much preferred to avoid direct confrontation on theological matters, feeling intellectually disadvantaged in such situations. In his later life he continued to read widely as time allowed; he favored the social sciences, stayed current with popular developments in theology, eventually dug deeply into business and "success" literature, and generally stayed away from fiction, hard science, and specialized monographs.

His seminary career, unpleasant at the time, gained nothing retrospectively. Many years later, while reeling from the critical attacks on *The Power of Positive Thinking*, he composed a recollection of these early experiences, perhaps hoping to put his current difficulties in context.[13] His seminary days offered a useful comparison. He remembered that when entering

> Theological School at Boston University my inferiority complex plagued me and contributed to deep unhappiness. I recall to this day how I mingled with the young ministerial students many of whom came from my own college feeling unworthy to be among them. Though I had graduated from a University as my father had not yet some deep inferiority made me feel that I was not qualified by culture and ability as were the others. To this day when in the company of a group of ministers that old feeling that I don't belong, that I am not their equal inhibits a full fellowship with them. I have studied, read and worked hard all my life to be a credit to the sacred profession of the ministry but the long memories [have] left an indelible imprint. In theology school we had a class in preaching. How I dreaded my turn. I watched the others as confident and self assured they preached before their fellows. The sermons were undeviating in style. All followed the same pattern. Similar expressions were used. They all seemed to know so well just how to compose and preach a sermon. So I tried to emulate and copy but was all too conscious that I was pathetically inadequate. To this day I can feel the hot flush of embarassment [*sic*] and the urge to get off alone to fight the battle of shrinking inferiority. Many an hour I paced under the trees on Boston Common struggling with myself to give it all up and go into some other field. But the Lord seemed definitely to be holding me to it. So I struggled on through three year[s] to which I look back as the [text obscured].[14]

His student church he recalled as an experience in contrasts, for there, he says,

> I forgot myself and lived in a perpetual thrill of praying, preaching, advising, and learning. On Sunday night in the glow of joy, after preaching to a church full of dedicated people I would return to Boston and as the train took me nearer to the school of theology a cold and icy chill drove the warmth from my soul. My conflicts returned, my inferiority reasserted itself and I froze up until the next Friday I went happily off to my church again.[15]

He continued,

> I do not blame the school of theology for my being as I was. But I do wonder if in the seemingly instinctive compulsion to force all ministers into the same

> pattern they are not now crushing the spirit of sensitive boys who are struggling through inner conflicts to be [sincere?] servants of Christ in the ministry.[16]

When Peale wrote about these seminary years more publicly in his autobiography, he recalled them as "bittersweet," acknowledging that on occasion his professors and peers had gone out of their way to be "kind" to him. But it was the "bitter" aspect he recalled, when, some ten years after graduation, as the minister of Marble Collegiate Church, he was invited back to speak in what he described as "the chilly atmosphere" at the school. He had recently been installed, in 1932, as the youngest senior minister of the distinguished Fifth Avenue church, and he may have taken some secret pride in knowing as he addressed the academic crowd that day that his meteoric rise to high position had come because of his rapport with the lay people, the "ordinary people" he idealized. But personal confidence failed to sustain him, and he was convinced that "they did not like me," that the speech went poorly, and that, therefore, he would never go back again.[17]

These private, painful recollections by America's foremost positive thinker cast the nature of his ministry in a fuller light. As a sociological description of marginality, they are reminiscent of W. E. B. DuBois' classic reference to the quality of "two-ness" for African Americans in white society: "One ever feels his two-ness; a Negro, an American . . .": or for Peale, as it were, "a New York minister, and a poor Methodist preacher's kid."[18] There was, to be sure, a significant difference between the racism that controlled DuBois' double life and the social marginality that created Peale's, yet there were also underlying similarities relating to marginality. Both refer to two worlds, two cultures: one marginal, accepting, and comforting; the other, mainstream, harsh, and judgmental. On a conscious level, it was this deep-seated feeling of social marginality that drove Peale to forge strong bonds with the "ordinary people" who became his constituency, creating a movement that found much of its identity in its oppositional stance, primarily to the so-called big shotism of a liberal intellectual elite.

Student Pastor, Berkeley, Rhode Island

The memoir left little doubt about where Peale found that "glow of joy" that gave meaning to his adult life, first as seminarian and then as successful preacher. For him, it was a fresh, sustaining experience every time he faced another sympathetic audience. Preaching became the centerpiece for his ministry, only later sharing that place with his interest in writing. One of his seminary courses in homiletics encouraged him to heed the advice of Phillips Brooks about confronting the congregation with truth and personality, which was advice similar in kind to the more homely admonition of his father to "be yourself."[19] It understandably took time to cultivate the appearance of naturalness.

Intent on his new calling, the young novice minister kept a diary in the early days of his career, intermittently recording his church appointments and his feelings. He noted that on April 3, 1922, he preached his first sermon, having been invited to conduct the service at the Methodist Church of Walpole, Massachusetts.[20] He commented to his diary:

> Today I preached my first sermons at Walpole, Mass. in the Methodist Church. Whether or no the audience derived any benefit from my services I do not know. I am sure, however, that a great spiritual joy flooded my own soul and a deep conception of the joy of the minister of Jesus first made itself felt in my consciousness.[21]

Licensed as a Methodist local preacher since 1919, he seemed confirmed that day in the certainty of his future calling. His text for his first sermon was his father's favorite, John 10:10: "I am come that they might have life, and that they might have it more abundantly." It soon became his own favorite as well. It served as the text for his initial sermon for each new ministry he began, and he invoked it so regularly thereafter that he once told an acquaintance that the text, with its promise of abundance, became "a sort of theme song" for his ministry.[22] After his maiden appearance in Walpole, he began to receive invitations to preach in other Methodist churches in New England during the summer of 1922, all of his sermons seemingly variations on the theme of personality he had been studying in seminary.

In his second year at seminary, in 1922, the Boston area Methodist bishop, Edwin H. Hughes, asked Peale if he had an interest in serving as student pastor of the church in Berkeley, Rhode Island. It would be his first pastorate, and as he indicated in his memoir, it became the kind of pleasant introduction to the work that allowed him to endure the rigors of seminary for the goal of the ministry. Although Berkeley was only a weekend charge, he was conscientious about his responsibilities, preaching twice on Sunday—including offering an altar call at the evening service—calling on members and getting involved in the life of the community.

His disaffection with experiences at the seminary allowed him to focus his energies on his ministry, establishing a pattern of single-minded commitment to work that would endure. It was a fevered, nineteenth-century evangelical work ethic that inevitably produced visible results. Uncertain and still untried in the pulpit, he found it difficult at Berkeley to follow his father's advice about preaching from the heart. Notes were necessary and sometimes a full manuscript. He began the practice of trying to find captivating titles for his sermons. One, titled "On the Homeward Way," required a detailed four-page outline to treat the Prodigal Son, asking listeners to think more about the ostensibly loyal brother—who remained at home, unsaved and envious of his wandering kin—than of the prodigal who returned. Another, at a Sunday evening service, concluded with a reading of the sentimental poem "Now I'm Coming Home," presumably to set the mood for an altar call.[23] The young Peale was obviously not much different in the pulpit from most other apprentice Methodist preachers, although his fierce commitment to work was exceptional.

The mill strike was on when Peale went to Berkeley, and it lasted through most of his tenure at the church. On occasion he spoke for the workers in negotiations with the owners in Providence. When his hometown newspaper in Findley, Ohio, reported on the activities of a native son, it credited Peale with winning a victory for the workers without alienating the owners. It also cited him for being instrumental in organizing a town meeting to block a proposed Dempsey-Wills boxing match in the area.[24] That same edition of the newspaper, appearing in the spring of Peale's last year in seminary, announced that he had accepted the call to a new Methodist church in Brooklyn, to be sponsored by wealthy St. Mark's congregation. He had been ordained to the ministry while at Berkeley, a date recorded in his autobiography as September 1922.[25] Thirty-five years later his own son, John, would serve the same church as student pastor.

A Star Is Born in Brooklyn

In moving to Brooklyn, New York, in 1924, Peale was able to enjoy the convergence of circumstances that would never again occur quite so fortuitously for him. He had moved into an area just beginning to open up as a community for New York businessmen and their families, at a time when hopes ran high and commerce and advertising provided the language of discourse. With energy and single-mindedness, he threw himself into the work.

The timing was perfect. A young seminary graduate, he was moving to an area in the midst of a population explosion as the optimism and ebullience of the Roaring Twenties shifted into higher gear. Methodist Bishop Luther B. Wilson had appointed him assistant at St. Mark's Church in Brooklyn, which was responsible for his salary, and minister in charge of the unsteady Flatlands congregation, whose membership Peale was expected to build up. He had doubled his salary in the move, from $1,200 at Berkeley to $2,500 in Brooklyn.

Flatlands was located on the edge of Jamaica Bay, and it remained a sparsely settled farming community until the Brooklyn Bridge and trolley cars opened up the more centrally located Flatbush. The Methodist congregation suffered attrition, and its property and tax expenses were assumed by the Brooklyn and Long Island Church Society. With the end of World War I and new transportation facilities, population began to push out from the city, resulting in an increase between 1921 and 1924 of about 30,000 people.[26] A group of Methodist Church leaders recognized the opportunity for church growth and arranged for the purchase of a lot on King's Highway to relocate the old Flatlands congregation, then meeting in its old frame building on Flatlands Avenue. The church membership was only 101, but the size of the Sunday school held promise for the future.

King's Highway, then just a muddy country road, was expected to be the site for a new, broad community church. Enthusiasm for the venture was running high when Peale arrived in the spring of 1924. Money for a new church was already pledged. St. Mark's Church and the Flatlands congregation each subscribed $10,000, the old church property produced $7,000 in equity, and the Methodist Board of Home Missions and Church Extension contributed over $16,000. Less than six months after his arrival in Flatlands, Peale participated in groundbreaking for the new church, on October 5, 1924, which appropriately turned out to be a beautiful fall day. The cornerstone was placed on December 21 the same year, and the new church was finally dedicated on September 20, 1925. Before the congregation could move into its facility, however, in the several months preceding the September opening, the membership so overflowed the old building that it was necessary to pitch a tent at the construction site. The press eagerly reported on the scene, the spectacular growth, and the lively new minister.

His tenure at the renamed King's Highway Methodist Church established a pattern for ministry Peale would continue throughout his career. His vision was fixed on the local church, primarily its growth, but also its services and program. He took full advantage of new information supplied to him by the culture, whether it was clues on advertising or opportunities for radio broadcasting. In his Brooklyn days, Peale's preoccupation with his work assumed an almost ascetic devotion: Unmarried and working in a new community, he lived in an apartment close to the church. In the three years he was at King's Highway, between 1924 and 1927, the church experienced phenomenal growth, increasing from just over a hundred members when he arrived to nearly 900 when he left, as a farming community was transformed into a suburban bedroom town almost overnight.

Peale experimented with whatever "new measures" came to his attention to recruit new members. Growth was not an unreasonable goal for a new church in a suburbanizing community; in the language of a recent student of the church growth phenomenon, Peale was "discipling" rather than "perfecting" members—that is, bringing them in in the anticipation that education would subsequently reveal to them the fuller implications of a richer, more self-conscious faith.[27] And as he had earlier adjusted to the needs of the local environment in Rhode Island, supporting the mill workers in their strike, so he did in Brooklyn by responding to the interests of the businessmen who influenced congregational life there.

As with many other liberal ministers of his day, he conceived a ministry within the culture rather than outside it and judging it. The motto on the cover of the Sunday bulletin made the point: "The Church—Life's Greatest

Investment." When new homes went up in the area, he secured from the power company the name of the owners, then called on them and invited them to church.[28] He studied other advertising techniques and applied them. During a "Buy a Brick" campaign for the new church building, a brass band preceded church members down the street as they solicited contributions for the new facility. He arranged with local radio station WBBC to have the weekly church service broadcast, and when the burgeoning congregation needed a place to hold its Easter worship in 1925, he secured the huge auditorium of the Marine Theater.

All of these innovations made good newspaper copy. Voluntary organizations for men, women, and children proliferated, and the Sunday school soon became the fourth largest in the city of Brooklyn. To keep up with the growth, the church hired two additional staff to assist the innovative young minister, and also paid professional singers to augment the voluntary choir. In realizing these ambitious programs, Peale cultivated the services of a group of laymen whose secular work exposed them to the New York City marketplace, and they eagerly related business techniques to church development. He nurtured bonds of familiarity with his congregation, who responded to his solicitousness with enthusiasm and warm regard: Where else in an increasingly anomic society could one turn for a sense of acceptance and familial comfort?

Winning members was the major focus of his ministry, but it was not the only one. In addition to program, the quality of the worship services claimed his attention. A perfectionist when it came to planning the Sunday church services, he had already abandoned any last vestiges of the informality of early Methodist evangelicalism. He adopted a more formal, even grand, style, evidenced by the music he chose and the gradual disappearance of altar calls. The choir might on occasion chant a prayer or sing a hymn in French or Spanish. In the midst of this solemn splendor, Peale would gradually unfold his seemingly boundless energy in his sermon: the contrast between liturgical formality and populist preaching doubtlessly startled listeners.

His message was already assuming the contours it would retain; it was a theologically liberal, inspirational talk that emphasized the transforming result of a relationship with Jesus and with the church. It was clearly a sermon about personal religion, with occasional references to Emerson already evident, although it also took heed of the social environment in which individuals had to live their lives. Impressed, enthusiastic, maybe even persuaded, people joined the church in groups of hundreds.

The clearest model for a "successful" ministry available to Peale at the time existed at neighboring Plymouth Church of the Pilgrims, its reputation secured not only locally but nationally by its colorful nineteenth-century pastor, Henry Ward Beecher. Beecher was a favorite of Peale's as he was of his father, and it was not lost on the younger Peale that part of Beecher's fame lay in his ability to preach practical sermons, always without notes. When

Beecher departed, the church chose a different type in Lyman Abbott, but when Abbott left, the congregation returned to a Beecher-type minister with Newell Dwight Hillis.

Hillis was ending a twenty-five year pastorate at the church just as Peale arrived in 1924. But his popularity offered lessons to the young minister on the ingredients necessary for a successful urban ministry. He had a charming, magnetic personality that suffused his exuberant, extempore sermons. He cultivated a national audience through the lecture circuit, giving his most requested speech, "Ruskin's Message to the Twentieth Century," to audiences across the country, saving a more specialized talk on "How Ability and Industry Increase the Worker's Wage and the Nation's Wealth" for a group of 2,500 newly arrived immigrants at Ellis Island.[29] Like Beecher, Hillis had drawn together the strands of liberal theology, practical Christianity, laissez-faire, and patriotism to create a message appealing to his sizable congregation, large portions of which drifted away—some perhaps to Peale's church—when he retired. His work was instructive to those city pastors who were eager to pattern their career on his, about the need to find just the right way to combine the ability of the preacher, the content of the message, and the needs of the people. Years later, Peale developed an even more personal interest in the ministry of Plymouth Church, when in 1941 the congregation called as its minister Wendell Fifield, brother of James Fifield, with whom Peale was associated in Spiritual Mobilization, a conservative organization promoting Christian individualism.

Peale's personal life apart from the church was virtually nonexistent during his years at King's Highway. He became a member of the Masonic Order, investing his participation in it with the same intensity he devoted to everything else. Ties of sentiment and emotion continued to bind him to his family: to his admiring brother Leonard he offered financial help with his education; and to his mother he tendered an occasional invitation to speak from his pulpit about her work with missions. He took pride in the fact that his father was finally recognized by the denomination by being appointed a Methodist District Superintendent, first in Findlay and then Delaware, Ohio. It was through his father's contacts with Adna Leonard, Methodist bishop of the central New York area, that he first learned that University Avenue Methodist Church in Syracuse would be looking for a new minister, and that Bishop Leonard regarded the younger Peale as an attractive candidate for the position.

Securing the Syracuse Spotlight

Peale was invited to preach at University Avenue Methodist Church in April 1927, and before the month was out he had been asked to become its new minister. On May 22 he preached his inaugural sermon, "The Glory of the Future," invoking the text from the tenth chapter of John's gospel, as his mother proudly looked on. The move to upstate New York, following a three-year pastorate in a community in the throes of urban growth, put him in a dramatically different environment, and consequently a very different parish. University Avenue Methodist Church was a settled congregation, located within easy walking distance of Syracuse University, an institution Methodist in origin if no longer in policy. The two institutions maintained a close, collegial relationship, so that to outsiders it appeared as if one were the extension of the other.

For someone with such unpleasant associations with higher education, it would seem that a university church would be an unlikely choice. Yet Peale betrayed no misgivings about his decision to leave a congregation of businesspeople for one of professors; only years later would he report on the anguish he suffered his first year or two in Syracuse trying to prepare "erudite" sermons for his intellectual audience. In fact, it was during his ministry in upstate New York, as the welfare of the nation and the health of the economy both sickened, that he consciously restyled his preaching, speaking more freely extempore, offering more practical advice, injecting fuller references to harmonial/mental science themes.

If a university parish seemed an odd choice for Peale, the city in which it was located was not. Syracuse was known for being snowy, ethnically diverse, and politically conservative, certainly more reminiscent of his Ohio background than Brooklyn had been. More blue collar than white collar, the city in 1927 had a metropolitan population of 200,000, sustained by a diversified industrial base, although the university was developing into the chief employer in the area. Syracuse University had been founded by the Methodist denomination in 1871, groundbreaking for its first building taking place on the same day that the first shovelful of dirt was turned for a new sanctuary for University Avenue Methodist Church. Their relatively similar origins, therefore, enhanced a connection based on denominational affiliation.

When Peale arrived at the church, the university was enrolling 5,000 students, a quarter of whom were Methodists. Because university policy required students to attend a church of their choice on Sundays, the proximity of University Avenue Methodist Church a half-mile down the "hill" from the campus guaranteed that the church would have a large student population. Faculty members and administrators were also in the congregation, making it possible for the church to have access to a rich pool of talent, particularly as it related to musical performers and guest lecturers.

The Central New York Methodist Conference paid the salary of the Methodist chaplain on campus—in 1928 it was the Reverend Webster Melcher—whose job description included planning and directing a program of activities for students. Because the university lacked a chapel at the time, these programs took place at University Avenue Methodist Church, producing yet another opportunity, on Sunday nights, for a large student gathering to occur at the church. In 1930 the university built Hendricks Chapel on the campus, and the heavy flow of student traffic down the hill to the church was much reduced.

The church building Peale inherited was a relatively new Gothic structure of grey stone, still heavily encumbered with construction debts. The original University Avenue Methodist Church (its name subsequently changed to University Methodist Church) had burned in a fire during a bitter winter storm in 1914, with the new one, because of rising building costs and the World War I, not completed until 1921. Large and resplendent, resembling a cathedral, the new building left the congregation with a debt of nearly $400,000, with $82,000 of it still owing when Peale arrived in 1927.[30]

The able young minister had already acquired a reputation that preceded him at the church. Only twenty-eight years old, he was known within the denomination for shepherding the remarkable growth of the Brooklyn church; a man, it was said, of unusual energy and "rare ability." Once again he streamlined his personal life to deal with the responsibilities of his work: In Brooklyn he had managed in a three-room apartment near the church; in Syracuse he stripped down still further, taking up residence at the Hotel Syracuse to eliminate all housekeeping problems. He recognized that he had two large issues to deal with. The debt was to be his first objective, since it was taxing the spirit of the congregation. And the second was to reorganize the student program to make it a more integral, more permanent, part of the church program. With 833 members, University Church was the second largest in the District, which helped explain why Peale received the handsome salary of $6,138 and expenses. Already he enjoyed a more visible status in the ministry than his father. Before plunging into his work, however, he took a vacation in Europe, whetting his appetite for a lifetime of travel.

The measures that had worked for him in Brooklyn were enlarged, adapted, and applied upstate. The growing penchant of urban Protestant ministers for efficiency and organization in their parishes was shared by Peale, who believed that the church had much to learn from the world of business and technology. He devoted a lot of his energy to developing advertising copy, and more to the planning of the Sunday service, its liturgy, and music, as well as his sermon. In the final analysis, his ministry was pulpit centered; the rest of the service, the formal liturgy, the stately music, the paid soloists, composed the elegant frame around his sermon.

His experience in Brooklyn had taught him that advertising could be effectively harnessed to the service of the church, and he remained an apt student of the subject in Syracuse. The ad for the church, now called only University Church, was the largest of the church advertisements in the local paper, appearing as well in the campus paper. Eye-catching titles for Peale's sermons appeared under such bold captions as, "Largest Vested Children's

Choir in the State," "A Great Church with a Great Purpose," "Come Early for a Seat," "Syracuse's Youngest Preacher," "The Greatest University Audience in America," and "Greatest Choir in Empire State."[31] Local radio station WSYR carried his weekly church service as the Angelus Hour. In 1928 the church board presented a five-point program to the congregation for its approval. It asked for a renewed consecration to God reflected in increased giving; an ambitious advertising program; an expanded musical program; a paid director of religious education; and a projected budget of $30,000. And it was Peale's preaching that was expected to drive the machinery to realize the goals, all predicated on boosting financial contributions from people whose modest incomes were largely derived from an educational institution.

His preaching, the source of his fame and subsequently his national celebrity, was redesigned in Syracuse. Peale later claimed that some of the inspiration for change came from a distinguished member of the congregation, Dean Hugh Tilroe, who administered the School of Speech at Syracuse University, and who became a close friend of the minister's. Tilroe, dedicated Methodist, stilled Peale's anxieties about creating highly stylized sermons for a university audience, advising him essentially as his father had, to preach from the heart. The professional teacher of speech gave Peale tips on public speaking, the range and projection needed in the cavernous sanctuary, and on the value of speaking without sermon notes or in a formal style. Peale, confessing he was intimidated by a presumed need to present "intellectual sermons" to his college audience, appreciated the advice and applied it; his style and content changed in ways that would endure. Animated, humorous, informal, he became an evangelist to the new age. Recalled by some parishioners some sixty years after his departure, he was remembered for his youthfulness and especially his enthusiasm.[32]

His sermons became increasingly contextual, picking up on the local culture, in particular its politics. And the local political issues that dominated his Syracuse ministry were the same ones that riveted the attention of the nation, namely, the implications of the 1928 campaign, which resurrected nativist themes around Al Smith's run for the presidency, and the sorry state of the nation's economic life. The general theme of his sermons, however, addressed matters of practical religion. He seemed to find the authority for this emphasis in essentially three sources: the mission of the church; the Personality of Jesus; and an awakened consciousness. This latter category was a relatively new addition to his message, though it obviously had a personal history with him. Importantly, it was the catalyst for the other two, for it was through an awakened consciousness that one experienced the personality of Jesus and made a commitment to the church.

Peale was convinced that the church had an important role to play in the culture, not least as the preserver of Protestant cultural hegemony. Although his faith in the institutional church would eventually waver, he told his Syracuse congregation that "Only the Christian Church can keep alive free institutions, the rights of man, and the sacredness of personality." He suggested euphemistically that the "better people" needed to take their religious responsibilities more seriously, keeping abreast of national and international developments without allowing the church to become "merely a current events institution."[33]

Peale's sermons spoke of the need for the church to work cooperatively with other agencies in the society, particularly the business community, to address the pressing issues of the day. Despite a sentimental longing for a pastoral past, Peale explained that "the city is and will continue to be the determining force in our American life." "Salina Street [the main business center]," he said, "will decide the fate of Christianity in Syracuse. Christianity will succeed or fail as it rises or falls on the Main Street in any city. The supreme test of its influence is to be found in the business and commercial life of our great urban centers."[34] The national economic slide then under way was making it painfully obvious, as churches suffered from reduced resources, that the business and financial community had greater impact on domestic priorities than the religious community. The church, he contended, needed to be an integral part of the culture, politically and economically, as well as morally and ethically.

Seminary had taught him that a relevant theology required cultural sensitivity and cultural involvement. Many of his professors would likely have been surprised, however, to see how he applied it.

His redesigned preaching offered fresh evidence of how the young pastor was learning to accommodate his seminary education to an increasingly personalized message. Sermons on the Personality of Jesus, for example, unaccountably provided the potential for political commentary. Too many people, he told his congregation, substituted a religion about Jesus for a religion of a personal Jesus.[35] From an intimate relationship with Jesus, said Peale, came emotional energy and

> Emotion breeds enthusiasm, and enthusiasm is that which is necessary to Christian world conquest. The need of the church today is for mystic or emotional contact with Jesus. This will create enthusiasm for world conquest. . . . The real need is a consciousness that Jesus is a living personality in the twentieth century.[36]

As he saw it, the demand of the hour for Protestant Christian America was to shore up the city upon the hill, and a reinvigorated, converted army of believers could help bring that about.

The concept of personality, as he derived it from seminary, became the focal point in Peale's developing message of personal religion. Individual transformation, he maintained, required a relationship with the Personality of Jesus. And it was through transformed individuals in the church that the world would be saved. "The hope of humanity," as well as the hope of the world, he said, "lies in positive and consecrated personality." Over the next five years or so, as he shaped his message, he alternately identified personality, as in the Personality of Jesus, with Mind, soul, or subconscious as the motivating force for individual transformation. The further he moved from Bowne's original theories of Personality, however, and the more he popularized his message, the fainter the original concepts became.

Eventually the theme of Personality was subsumed within the concept of mind [though not yet Mind] in his sermons, resulting in preaching that was increasingly focused on practical concerns. It marked the beginning of a shift in his ministry along the religious science continuum, a tentative shift from a mystical Methodism to a form of New Thought. In a sermon at University Church near the close of the decade, he introduced his congregation to two positive thinkers: one was the apostle Peter and the other was the Arctic explorer Richard Byrd. The goals they achieved, Peale pointed out, driven by inner qualities potentially available to everyone, suggested that heroic accomplishments came from "harmonizing life with the spirit of Christ."[37] Their feats, he maintained, were not extraordinary, for "given a normal intelligence, an individual can make of himself about what he wants to be." What these goals should be was outlined in Jesus' Sermon on the Mount, which he called a guide for a "practical program for personality building."[38] Peale's sermons were increasingly focused on problems of daily living, but were not yet much concerned with technique.[39]

No single epiphanic moment marked a change in his message, a willingness to identify personality with Mind in the tradition of mental science healing. During pastoral visits in the homes of Syracuse church members in these years, Peale claimed to notice a proliferation of the literature of Christian Science and other metaphysical healing groups.[40] But there was no direct causal link. The demands of the local culture, for growth, to pay bills, urged him to increasing sensitivity to listeners' needs as he crafted his message. The national preoccupation with success, and as the Depression approached, just getting by, encouraged it. He was also personally driven to succeed, to "make something of himself," to make the church grow, and he appreciated the popular appeal of a message of practical Christianity.

He also believed, as his seminary professors did, that theology and religion had political relevance, only in his case it was in terms of a conservative civil religion. For the first three years of his ministry in Syracuse, the fight over Prohibition consumed much of his energy and politicized his ministry, until ultimately his Prohibition sympathies were transformed into anti-New Dealism. The Prohibition struggle had been a combined Peale family fight, and Norman continued to wage the battle in central New York. He had plenty of company. The local Central New York Methodist Conference, headed by Bishop Leonard, sent a message of congratulations in 1928 to the flamboyant Prohibitionist crusader, Bishop Cannon, undoubtedly wishing years later it could be retracted when Cannon's career was scarred by scandal.

Peale's highly charged political instincts kept him right in the thick of the fray locally. Al Smith's defeat in 1928 had temporarily assuaged "Dry" fears, but as the election of 1932 approached, given New York State's reluctance to enforce the Eighteenth Amendment, it appeared certain that another test of "Dry" strength was imminent. Peale went on the offensive against the controlling "Wet" minority within the Onondaga County Republican Party and gained arresting headlines in the local paper: "Peale Flays Party Chiefs of Onondaga." In a more open denunciation of the same group later that year in a speech before the WCTU Convention meeting in Syracuse, he accused "Our Wet and vicious newspapers" of undermining Prohibition and called for the election of an alternate set of Republican candidates on the Law Preservation Ticket. During the summer of 1931 he had traveled with his new wife throughout the West, speaking at Methodist conferences, touring for a time in Mexico, and returning to Syracuse almost persuaded the country would stay dry, because the West supported Hoover and consequently Prohibition.

Few partisans made any effort to disguise the tribal nature of the Prohibition struggle, and surely not Peale. He advertised a sermon in 1929 that made it quite explicit: "The Pope Looks at Protestantism—Is It Exhausted? Will We All Be Catholics?"[41] His sentiments were in line with those of mainstream Protestantism, which, as Robert M. Miller has pointed out, endorsed Hoover in 1928—despite Smith's being more fiscally conservative—with even the liberal *Christian Century* joining in. A whispering campaign alleging a Catholic conspiracy provided the emotional undertone to keep the struggle alive.[42] For Peale as for many other Protestants, the Victorian evangelical-Whig synthesis, the belief in the identity between Protestant values and bourgeois culture, still endured, however precariously. His preference was for a personal religion of practical Christianity, which depended on sustaining the synthesis through individuals, yet through the 1930s he continued to recognize the institutional church as the agency for implementing the will of its members.

His Prohibition passions and conservative tendencies did not preclude his attending a 1931 denominational conference convened by the liberal Methodist Bishop Francis J. McConnell, under the auspices of the Board of Foreign Missions. Meeting for ten days in Peale's old collegial bailiwick of Delaware, Ohio, the conference agenda expressed the progressive intentions of Bishop McConnell and his conference colleague, Professor Harry Ward of Union Seminary, both of whom also served as catalysts in the controversial Methodist Federation for Social Service. The Ohio conveners obviously intended their session to be another means for informing a report on the State of the Church to be presented at the 1932 General Conference. The delegates predictably endorsed McConnell's liberal agenda, supporting birth control, prison reform, desegregation, unemployment insurance, old age pensions, government regulation of holding companies and other monopolistic business practices, and disarmament.[43] Since the Washington Naval

Conference of 1921, reduction in armaments had become a compelling issue, especially to the major naval powers of the United States, Britain, and Japan, and Peale along with his own home bishop, the basically conservative Adna Leonard, had been an outspoken proponent of disarmament.

His support for disarmament fell short of isolationism. In a talk to students, he decried the "psychology of war," arguing that the church had the responsibility to teach the world to abolish war.[44] To a Methodist conference in Idaho in 1931, he scored the press for its "pernicious influence against disarmament" that was resulting in "harrowing the nation into war." Furthermore, he told that same group, "something is fundamentally wrong with the roots of the social order" when maldistribution of resources lead to unemployment rates between five and eight million, creating "a tragedy of plenty."[45] Still close to his seminary experience, Peale continued to cling to the hope that the institutional church might in fact awaken the nation to deal responsibly with social needs.

Coping with the Crash

The tragedy of the Crash of 1929 occurred midway through Peale's tenure at University Church and soon had a devastating effect on a congregation struggling to get out from under the burden of heavy indebtedness. Soon after coming to Syracuse, the new minister had planned a financial campaign to reduce the mortgage, designed to raise $60,000 over five years. Convinced that the major group in his congregation—the professors, students, and university employees—were loyal but lean sources of support, he took his campaign to the businessmen in the church and the community. In a prayer meeting with one of the church stalwarts, a man who had lost and remade several fortunes in the wholesale grocery business, he secured a pledge for $5,000. The generous contributor, Harlowe B. Andrews, intended his gift as an antidote to the "same old negative attitudes" that were undermining the budgetary goal.[46] Reporting on the event years later, Peale said, "I always wanted his [Andrews'] secret of making money."[47] Like early Progressives, including his own father, Peale believed that the business community was composed of decent Christian men who needed only encouragement to respect high moral principles, a view that also smoothed his interactions with the men with whom he regularly lunched at the Syracuse Rotary Club.

His financial campaign succeeded in raising the $60,000 in pledges, and the District Methodist Conference commended him in 1928 for a "signally successful year," for meeting the budgetary goal and for adding 316 new members to the church. Although some of the membership gain was attributable to a new policy of affiliated membership for students, it was still an impressive increase.[48] The church budget reflected the unsteady health of the national economy; just under $25,000 in 1927, it rose steadily to $27,500 in 1928, $34,000 in 1929, $36,000 in 1930, and then peaked at $37,300 in 1931 before declining precipitously in 1932 to $23,098, often called the trough of the Depression. Peale's own ministerial support was cut, as it was for over half the clergy in the conference, from $9,700 in 1931 to $7,300 in 1932.[49]

The membership growth at the church was considerably more noteworthy than its budget. The university community, short on large pledges, contributed dramatic congregational growth: following the 316 new members added in 1928, another 345 joined the next year, making the church the largest in the Conference and the second largest in Syracuse. Each succeeding year, hundreds more joined, so that after factoring in losses due to deaths, transfers, and some attrition, the membership had nearly doubled by the end of his six-year pastorate in 1932, going from slightly over 800 to 1,600. The measures that explained the extraordinary growth were the same ones that had worked in Brooklyn: Peale's preaching, clever advertising, and an active program planned with care and creativity. In addition to the usual parish activities, like an Epworth League and a Ladies Aid Society, there were specialized offerings, ranging from a student church, an organization of "preachers' kids," to a spirited men's club, basketball for nurses, and lectures by international experts, along with concerts by distinguished musicians. Peale's own reputation kept pace with the growth, as his audience swelled beyond the local parish when he signed on with a lecture bureau, which paid him for making guest appearances. The newspaper reported that he regularly preached to over a thousand at University Church and ventured the opinion that he was one of the most eloquent and attractive speakers in the state. He possessed an uncanny ability to understand the existential moment, to sense an audience and sculpt his message to its sensibilities; he had the intuitive skills of a populist, now honed by experience. Audiences didn't like him; they loved him.

Ruth Stafford: Team Partner

Peale's enduring legacy from Syracuse was Ruth Stafford, the university student who in 1930 became his wife. Theirs eventually became a relationship of extraordinary compatibility, lasting over sixty years. Though Norman would later write many books and articles containing allusions to his private life, he never disclosed the secrets he and Ruth learned about making their marriage so unusual and seemingly so rewarding. Each cultivated patience, forbearance, and affection in responding to the other's idiosyncrasies: in Ruth's case, for Norman's phlegmatic personality, and, in midlife, long absences; in Norman's, for Ruth's managerial style with its relentless attention to detail.

When he first met her in 1928, Ruth Stafford was a senior at the university, a mathematics major whose home was Detroit. Similar personal histories doubtless facilitated their relationship. Like Norman, Ruth was a Methodist preacher's kid from the Midwest, born in Fonda, Iowa, and was one of three children. At the university she lived at Alpha Phi sorority house, where she roomed with the daughter of Norman's local Methodist bishop. Petite, blond-haired, and according to one of her Sunday school students, having "big blue eyes," she was an attractive coed, a good student, and a proper Methodist. Although she struggled with her childhood experiences in a parsonage, as Norman did, and vowed never to marry a clergyman herself, she took an active role in Methodist functions on campus. When University Church created a student church, she was elected its president, a position that made it necessary for her to meet the minister. Norman and Ruth developed different memories of that first meeting, although both recalled its taking place, appropriately, in the church sanctuary, with late afternoon sun streaming through the stained glass windows. Ruth remembered that he held on "just a trifle longer than necessary" as they shook hands, while Norman's impression was that the beautiful young woman had come to wait for one of the other students.[50]

Ruth soon became an active participant in the church, worshipping with the congregation, teaching Sunday school, and presiding over the student church. The minister and the student leader met at committee gatherings and then over quiet dinners alone. Norman was thirty, Ruth nearly eight years younger, and he appeared to have more serious romantic interests sooner than she. She was probably his first real amorous interest; except for his brother Leonard's passing reference to "hearing about your affair in college," the Peale record is dramatically silent about matters of the heart.[51] The two continued to see each other during Ruth's senior year at the university and then for the next two years, as she worked as a math teacher at the local Central High School in order to be near him. Finally engaged in the spring of 1930, they were married in University Church before an overflow crowd on June 20, 1930. Three Methodist clergymen officiated at the ceremony: Norman's father, of course, along with Methodist Bishop Leonard and Chancellor Charles Flint of Syracuse University. When all the events of the day were finally tidied up—the 4 o'clock Friday afternoon wedding and the small party afterwards—the couple took off for a night at the Inn in Cooperstown. The following day Norman gave the address at Cazenovia Seminary, some twenty miles from Syracuse, where his youngest brother Leonard was a member of the graduating class. Only then did they leave for a honeymoon of several days in the Adirondacks, in a cottage made available to them by Syracuse friends.

After they returned to Syracuse, Norman moved out of the hotel and the couple took an apartment. Norman returned to his work, and Ruth turned her considerable talents to what would become her life's work—organizing the family's interests to enable her husband to devote himself single-mindedly to his work. She appeared to find not only identity but satisfaction in the arrangement, as many political wives had done. Ruth had full control of the household, although Norman had as much interest as she in a tasteful, well-ordered home; sixty years later, one of Ruth's former Sunday school students would recall her own wide-eyed amazement at the sleek, modern, black couch with satiny stripes Ruth ordered for the apartment.[52] Ruth organized all of the details of their lives, including Norman's wardrobe, leaving him free for his preaching and pastoral work. Despite marriage and then family, he would enjoy the great luxury of never being distracted by the ordinary details of living from his singular commitment to his work. He consistently referred to Ruth as his partner, explaining that theirs was a team ministry, as in fact it was. Similar models could be found in the marriages of other well-known evangelists. She was his devoted and only confidante, the person to whom he turned for information, comfort, and advice, an equal architect of his work. And once their children were grown, she entered into the work herself, organizing a committee to publish and circulate Norman's sermons, which eventually outgrew itself and became another Peale enterprise, the Peale Center for Christian Living, which she headed.

A Call to New York

Peale's reputation as a popular preacher with the ability to facilitate church growth could not long be contained in Syracuse. After declining a number of invitations over the preceding several years, in March 1932 he agreed to be a guest preacher at Marble Collegiate Church in New York City, a church seeking a replacement for the Reverend Daniel Poling, who had resigned two years earlier. In marked contrast to University Church, where it was necessary to arrive early to get a seat, Marble Church had a tiny congregation. But also unlike University Church, it had a sizable endowment; through its ties to the Collegiate Church system—part of the Reformed Church in America—it was rumored to be the third largest property owner in New York City. The committee, not surprisingly, was awed by Peale's performance and expressed its interest in having him become their next minister. He explained instead of responding. For the coming month of May he would be preaching at First Methodist Church, Los Angeles, where the minister was retiring soon, and from which he had reason to suspect a call would be forthcoming. At the end of his stay in California, he did indeed receive an offer to become the next minister, and he was then confronted with the need to decide between two impressive, but radically different opportunities.

For a month, he and Ruth considered the two options. First Methodist Church, Los Angeles, was the largest in the denomination, and with over seven thousand members, one of the largest in the country. Its evangelical program was familiar and acceptable to him, as was the climate. Marble

Church, New York, with a lineage going back to 1628, offered historic prestige, Fifth Avenue status, and the considerable financial resources of the Collegiate Church system. Marble's committee presented Peale with an impressive personal package, which limited his responsibilities to preaching and administration, allowed time during the week for other speaking engagements, provided long summer vacations, an attractive salary, and an appointment for life. Working against a decision to go to Marble was the fact that, located in a part of the city that had passed its prime and having been without a settled minister since Daniel Poling resigned to fight Prohibition full time, it had become a congregation of only a few hundred active members. It was also part of the Reformed Church in America, which would mean leaving Methodism. The impasse over which offer to accept was finally ended, Peale later recalled, as an answer to prayer. It was probably also a choice that was psychologically comfortable, conforming as it did to a personal credo Peale had developed about choosing the lesser of two options: a poorer situation could only improve with the application of a bit of hard work, whereas the future of a successful operation could be extremely uncertain.[53]

In the removal to New York, there were more substantial issues to resolve than finding suitable housing in the City. In accepting the call to Marble, Peale was required to transfer his ordination from the Methodist Church to the Reformed Church in America, theologically a shift from Arminianism to Calvinism. As he thought about it retrospectively he observed in an unusual understated fashion: "While, of course, I have never been bound by denominational ties, yet I did feel a twinge as I left the church of my fathers."[54] In many ways, he was never to leave Methodism, for unable to own the hard determinism of Calvin, he created a new style of ministry, its synthetic message shaped by his Methodist heritage to be sure, but now open to the language of Calvinist orthodoxy and the harmonial optimism of metaphysical spirituality. His symbolic break with Methodism, coming just as his personal fortunes were rising while the nation's plummeted, released him to create a ministry that cut across denominational lines. It eventually became a movement of which he was the sole designer.

On October 2, 1932, in the unusually crowded sanctuary of Marble Collegiate Church, he was installed in a solemn ceremony as its new minister. Dr. Charles E. Jefferson, honorary minister of Broadway Tabernacle in New York City, preached the sermon. Clifford Peale, now minister of the First Methodist Church of Columbus, Ohio, had the more personally rewarding task of addressing his son through the ritual charge to the minister. His father lightly observed that Norman had come into the world yelling and was still yelling, adding more poignantly, "You have never disappointed me, and you will not disappoint me now."[55]

Peale had survived the troubled, tribal twenties unscathed. His new neighbor, Harry Emerson Fosdick, a victim of the fundamentalist-modernist controversy as well as a recent convert to the need for counseling and personal healing in ministry, could not say as much. Peale believed that in his own work he trod the middle of the road, and that his move to New York in 1932 was timely and portentous. Surely many of his colleagues would have disagreed, viewing a new ministry to a dwindling congregation during the darkest days of the Depression as evidence of questionable judgment. Peale's optimism was linked to two factors—one personal and the other professional—though they were certainly related. Personally, Marble Church gave him financial freedom and a temporal autonomy that allowed him to carve out his own identity, to experiment with new approaches, and to attempt to find a national visibility. And professionally, equipped with experience gained in managing two different models of church growth, he was prepared to develop an innovative ministry around a message of practical Christianity. He hoped it would help counteract the personal tragedies produced by the Depression, rebuild Marble Church, enable him to reach a wider constituency, and not the least significant, fight the New Deal.

Notes

1. Sydney Ahlstrom, *A Religious History of the American People* (New Haven, CT: Yale University Press, 1972), pp. 896–98; Robert T. Handy, *A Christian America* (New York: Oxford University Press, 1984), pp. 161–65.
2. Handy, *Christian America*, p. 169.
3. Martin Marty, *The New Shape of American Religion* (New York: Harper & Row, 1958).
4. Robert Moats Miller, *How Shall They Hear without a Preacher: The Life of Ernest Fremont Tittle* (Chapel Hill: University of North Carolina Press), pp. 38, 39.
5. Transcript of NVP, Boston University School of Theology; NVP Ms. Coll. Although he claimed to find his two courses in practical preaching extremely painful, he managed an A and a B in them.
6. NVP, Boston University School of Theology, Department of Field Work Report, 1924; NVP Ms. Coll.
7. Ibid.
8. Miller, *How Shall They Hear without a Preacher*, p. 485n.
9. Borden Parker Bowne, *Kant and Spencer: A Critical Exposition* (Port Washington; NY: Kennikat Press, 1967), p. 10. These are lectures published after Bowne's death in 1910.
10. Walter G. Muelder, *The Ethical Edge of Christian Theology* (New York: Edwin Mellen Press, 1983), pp. 1–42. Muelder also studied at Boston University School of Theology in the twenties and here discusses his experiences studying with Knudson and Brightman. See also Ahlstrom, *Religious History*, p. 777.
11. Muelder, *The Ethical Edge*, pp. 7, 8, 72.
12. In file of student papers and class notes, NVP Ms. Coll.
13. See Chap. 1, n22. The material quoted here is taken from the same extended recollection cited in Chapter 1.

14. Ibid. A portion of this handwritten, seven-page reminiscence, quoted in Chapter 1, noted his feeling about his home life, and the lack of preferment for his father by the Methodist hierarchy.

15. Ibid.

16. Ibid.

17. Norman Vincent Peale, *The True Joy of Positive Living* (New York: Morrow, 1984), p. 89.

18. W. E. B. DuBois, *Souls of Black Folk* (Greenwich, CT: Fawcett Publisher, 1961).

19. NVP, Seminary notes, 1923; NVP Ms. Coll., Box 827.

20. He would later commemorate April 3, 1921 as the anniversary of his first sermon. Peale entered seminary in September 1921 and in April of that year would have been still engaged as a newspaperman in Detroit. It is possible that he left Detroit in the spring of 1921 to be in New England for a week or longer, and thus could have preached the sermon on the date he remembered. It seems more likely that it was 1922, when he was already in seminary.

21. NVP Diary, April 3 (no year); NVP Ms. Coll.

22. Letter of NVP to Alton C. Roberts, Jan. 4, 1961; NVP Ms. Coll., Box 342,

23. NVP Sermon Ms., "On the Homeward Road," Jan. 21, 1923; NVP Ms. Coll.

24. Findley, Ohio, *Morning Republican*, April 17, 1924.

25. There is some doubt about this date. On March 11, 1923, the Quarterly Conference of the New England South (Methodist) Conference approved a request from Berkeley Methodist Episcopal Church for the ordination of Norman Vincent Peale as deacon.

26. History of King's Highway Methodist Episcopal Church, pamphlet; NVP Ms. Coll.

27. A quote from Donald McGavran, *Understanding Church Growth* (Grand Rapids; MI: Eerdmans Publishing Company, 1970), in an article by Robert A. Evans, "Recovering the Church's Transforming Middle: Theological Reflections on the Balance between Faithfulness and Effectiveness." In Dean R. Hoge and David A. Roozen (eds.), *Understanding Church Growth and Decline: 1950–1978* (New York: Pilgrim Press, 1979), pp. 302ff.

28. Interview with NVP, February 24, 1983, Pawling, NY.

29. Plymouth Church of the Pilgrims, *A Church in History* (Brooklyn; NY: Stefan Salter, 1949).

30. Nelson M. Blake, *History of United University Methodist Church* (privately printed, 1970); W. Freeman Galpin, *Syracuse University: The Growing Years* (Syracuse; NY: Syracuse University Press, 1960), Vol. II, appendix.

31. Selected ads from the Syracuse *Post Standard* and the Syracuse University *Daily Orange*; some in Syracuse University Library, some in NVP Ms. Coll.

32. Conversations with Frieda Kirkley, April 1, and April 3, 1987, Syracuse, New York. Other people also present in the church hall at the time contributed their recollections to the conversation.

33. Syracuse *Post Standard*, September 20, 1931.

34. Ibid., May 21, 1928; May 23, 1931.

35. Ibid., July 22, 1929.

36. Ibid., July 2, 1928.

37. Ibid., December 3, 1928; December 9, 1929.

38. Ibid., May 25, 1931.

39. Herkimer, New York, *Evening Telegram*, June 25, 1930; also Church bulletin files, Archives, University United Methodist Church, Syracuse, NY (formerly University Methodist Church). Some of the reported sermon titles were: "How to Develop Inner Strength," "How to Be a Better Christian," and "Overcoming the World."

40. Interview with NVP, December 29, 1980, Pawling, NY.

41. Syracuse *Post Standard*, December 1929.

42. Robert Moats Miller, *American Protestantism and Social Issues* (Chapel Hill: University of North Carolina Press, 1958), pp. 50–58.

43. Reported in the Syracuse *Herald*, September 8, 1931.

44. NVP speech, Syracuse, New York, Apr. 14, 1928; NVP Ms. Coll.

45. News report of speech in Buhl, Idaho, Aug. 23, 1931; NVP Ms. Coll.

46. Peale, *True Joy*, p. 116. This meeting with Andrews has frequently been retold by Peale in sermons and other writings.

47. Syracuse *Herald Journal*, November 29, 1981.

48. Central New York Methodist Conference, *Annual Yearbook*, 1928.

49. Central New York Conference, *Annual Yearbook*, 1927, 1928, 1929, 1930, 1931, 1932, 1933.

50. Letter of H. Halstead of F. Witmeyer, April 18, 1984, University United Methodist Church, Syracuse, archives. Peale recalled for his autobiography the setting for this first meeting in a similar way, while indicating that he thought Ruth had come to the church to wait for one of the students on the committee. Peale, *True Joy*, p. 110. Ruth has also told her recollection of the meeting on many occasions.

51. Letter of Leonard Peale to NVP, Nov. 2, 1932; NVP Ms. Coll. There was also a rumor circulated at University Church that a young woman in the congregation who was driven home from church meetings on a couple of occasions by the minister had a special relationship with him—she was the author of the rumor—but Peale quickly scotched the story. Given his usual shyness in small groups, it seems likely that he would feel awkward asking a woman for a date, especially someone from within his own congregation.

52. Conversation with Frieda Kirkley, University United Methodist Church, Syracuse, April 1, 1987.

53. Peale has related this bit of personal philosophy in many places—in books, sermons, talks, interviews.

54. Peale, *True Joy*, p. 130.

55. Ibid., p. 131.

CHAPTER 3

Drawing Down the Lightning 1932–1942

By leading their thought [the orator] leads their will, and can make them do gladly what an hour ago they would not believe that they could be led to do at all: he makes them glad or angry or penitent at his pleasure; of enemies makes friends, and fills desponding men with hope and joy. . . . Eloquence is *the power to translate a truth into language perfectly intelligible to the person to whom you speak.*

Emerson
Essay on Eloquence
1867

"Not yet 34 and the Marvel Preacher of Methodism." The advertisement in the Los Angeles *Examiner* for April 30, 1932, announcing Norman Peale's month-long candidacy at First Methodist Church in the city, highlighted the quality that subsequently earned him job offers from two prestigious congregations. His first appearance in *Who's Who* in 1930, when he was reportedly the youngest entry, had been essentially a statement about what his gifted preaching had produced: the growth of King's Highway Methodist Church from 40 to 900 members in three years and that of University Church from 800 to 1,600 in six years. After two agonizing months of indecision, Peale finally accepted the offer from Marble Collegiate Church, convinced that he could accomplish in Manhattan the same magic he had worked in his two previous charges.

The general pattern Peale established for his work during the thirties persisted throughout his ministry, and, like his message of spiritual wholeness, was simple and straightforward. Everything was based on preaching or public performance. Marble Church had been his choice, he said, because the New York pulpit offered "the opportunity to make preaching the one soul [*sic*] work." He had seen preaching produce growth, and he began his Marble ministry with almost millennial expectations for a revitalized Protestant Church, not only in his own congregation and in New York City but across the nation. Such an institution might be capable of recapturing the old Protestant cultural hegemony.

Effective preaching required that the content be accessible and integrated with an arresting style, and Peale groomed both to appear as a unified, artful presentation. By the middle of the decade the content of his message had achieved the kind of unambiguous, uncomplicated quality it would retain. His theme was practical Christianity, applied to the realities of day-to-day living. In a sense, the credo of his work was the Emersonian observation, "The thought is the ancestor to the deed." What that came to mean was that through prayer, self-examination, and surrender, individuals could become channels of divine energy, empowered to realize their goals. A primary objective of the newly awakened individual should be personal self-control, the ability to gain control over one's own life and the crises that intruded on them.

Yet when he arrived in New York City in 1932, the local environment provided stark commentary on how Depression blight had devastated the lives of people, economic casualties quite unable to control their own lives. It was with the needs of these victims of socioeconomic collapse in mind, Peale said, that he began a multifaceted program of practical Christianity, creating a message and an agenda that came to have, seemingly unintentionally, gender-specific relevance. At the end of the decade, he launched a specialized ministry to businessmen. Earlier, he inaugurated an Emmanuel-style psychospiritual healing clinic at Marble Church, serving a largely female clientele. Not incidentally, the financial support from the business community helped fund the various projects he assembled in his friendly, surprisingly unstuffy church, including the clinic budget. His hope was that restored, healthy, energized men and women would help vitalize a church that sprang from "the people," one largely oblivious to institutional and creedal boundaries.

During the decade he tried out his message in the church, in the clinic, and in his ministry to businessmen. He also attempted to shape a nondenominational national constituency by utilizing the best resources of technology and communication, experimenting with what his younger successors would fashion into the electronic church. He wrote books and a newspaper column, took railroad sleepers to distant audiences in scattered cities, and even tried to establish at home an extradenominational mental science congregation patterned after the successful program of religious science preacher Emmet Fox. Those seeds planted, he could then turn his attention to plowing back the New Deal.

Crafting a New Ministry

Peale went to Marble Church with few illusions. That he decided on a dwindling, wealthy Reformed Church congregation in the city known as the graveyard of ministers as against a large, active West Coast parish in his own denomination—after prolonged consideration of both—indicates that he recognized a fit between his interests and what Marble had to offer. When first wooed by Marble with an invitation to preach in its vacant pulpit in February 1932, he accepted, he told his parents, because "I need the money"—the $125 honorarium. He dismissed the prospect of a serious candidacy, he assured them, because it would mean facing the question of whether or not to leave the Methodist Church.[1] In March, on the eve of his departure to speak in New York, he was still writing to his father to thank him for his help in gaining the attention of the Los Angeles church, a call from which would be a sure sign that "God was directing it."[2] But by the time the Marble Church search committee visited him in Syracuse he had evidently changed his mind, for he told them he "would go anywhere that a big opportunity presented itself." To his anxious, hypertensive mother, he counseled, "My dear, dear Mother let peace come into your soul for it will give you the strength you need."[3]

Marble's offer sounded like a "most impressive proposition": ability to draw from a generous endowment fund and a clear delimitation of clerical duties. As the nation slipped into the depths of the Depression, Peale was acutely aware of the financial toll it was taking on church budgets like his own at University Methodist. Marble Church had some financial problems, but its ability to share in the resources of the property-rich Collegiate system created an anchor of security in the midst of general economic turmoil.[4] This financial health was reflected in the church's ability to provide Peale with a significantly larger salary than he could have anticipated in California: $15,000 versus $9,000 in Los Angeles.

The really decisive factor in his choosing New York, however, was the nature of the job description: At Marble, with lifetime tenure and an annual three-month vacation, his only requirement was to preach and function as the chief administrator. Here was surely a great opportunity for an ambitious young man to create a personalized ministry. When Peale finally made the painful disclosure of his decision to his local bishop, Adna Leonard, whom he loved "like a father," he emphasized his intention of remaining faithful to Methodism: "I do not go" to Marble, he wrote, "as one who goes to a denomination. I go in my own mind to a particular church to preach the gospel of Christ. You may be sure that my preaching will have to the very end the strong evangelical note given me by my Methodist heritage."[5]

Peale lived up to his promise to Bishop Leonard during his pastorate of over fifty years at Marble. His theology and preaching, though marked by his own handiwork, were more clearly in line with the evangelical Methodism of his youth than with the Calvinism of his adopted denomination. In any case, denominationalism—like ecumenism—was irrelevant to him, as presumably it was to the thousands who would regularly join Marble in the future. They were attracted to Marble by Peale, and they came from all the major denominations and no denomination: Over the years a majority, usually 75 to 80 percent, came from mainline Protestant denominations, with about 10 percent coming from Roman Catholicism and another 10 percent from a variety of other belief groups and sects.[6]

His new colleagues in the Reformed Church in America were somewhat less sanguine about his choice of a new spiritual home. In a small denomination like the Reformed Church, which in 1932 had a fairly stable membership of slightly under 200,000, the practice of "going outside" to secure a pulpit star for one of the few prestigious appointments in the church smacked of disloyalty: There were very few choice positions to which talented Reformed Church seminary graduates could aspire.[7] That Peale had to undergo years of scrutiny before he was accepted by his new denomination, and even then to receive less than total acceptance, added to his natural suspicion of denominationalism.

The Reformed Church had a long and proud tradition, and denominational stalwarts would surely have preferred to have an internal candidate invited to Marble's pulpit. The Collegiate Church in New York, a collective that included at various times four to five city churches—of which Marble was one—and several smaller congregations in the boroughs, prided itself on being the oldest Protestant church with a continuous ministry in the United States. Founded in 1628, it claimed Peter Minuit and Peter Stuyvesant as early members. Marble Church had been built as one of the Collegiate churches in 1854, beyond the Twenty-third Street city limits at the muddy, rutted intersection of Fifth Avenue and Twenty-ninth Street. Constructed of quarried marble from Hastings-on-Hudson, New York, it was a stately, impressive building of Romanesque design. The administrative structure of the Collegiate system, with life-tenured ministers and nearly life tenure for appointed elders and deacons on the lay board of the ruling Consistory, preserved a conservative policy, particularly in matters political and financial.

But tradition had not spared Marble the ravages of urban change. The tall spire of the handsome church was dwarfed by surrounding commercial buildings, a local residential neighborhood already a thing of the past at the time of Peale's arrival. The elegant sanctuary with seating capacity for 1,500 boasted Tiffany windows, red carpeting, gold-flecked maroon walls, and swinging pew doors; but since Dr. Poling's departure in 1929 it had been the home of a congregation averaging only about 200. There had even been a vote to consider closing the church. Financially, the picture turned out to be not much brighter. The amount of money raised by the local congregation was just over $12,000, subscribed by 264 members, augmented by a single gift of $8,000 from one member, making for a total annual budget of $20,000. The collective Collegiate Church resources paid Peale's salary and other operating expenses, but Marble was far from a financial Eden. Money for new programs and activities would have to come from an expanded membership and revitalized church life. Years later, as he looked back on his initiation to his Marble ministry, Peale enjoyed recalling a particular wind-swept night when he stood on the church steps with his mother. In an emotional moment, she leaned against the marble pillars and said, "[This is a strong church,] keep it strong, my son." At the time it seemed like a difficult promise to keep.

He confessed that the main attraction of the New York pulpit was the ability to concentrate on preaching. But he also said that it was his intention to "begin immediately a spiritual clinic," with daily office hours open to any "needy" person who sought his help. He was predictably besieged by those from within the church and without who had suffered from the exactions of the Depression.

A Dismal Start

On the night of Peale's installation as minister, October 2, 1932, Fifth Avenue was gaily decorated with lights, flags, and bunting to celebrate its twenty-fifth anniversary as a "street of commerce and art," but its shops, like the church, lacked patrons. The commemoration was as much a display of wishful thinking as was Peale's installation.[8] The church was packed to standing room capacity for the installation, a phenomenon not to be repeated, except for special events, for another ten years.

Hoping to duplicate the magical transformations he had worked in Brooklyn and Syracuse in an environment now less kind, Peale drove himself harder, made his personal life still leaner. With Ruth no longer teaching and having as yet no children to care for, she managed all the details of personal living. She engineered their relocation from a Gramercy Park hotel to a Fifth Avenue apartment near the church, secured for them by the Consistory. Peale forfeited all hobbies, even the limited golf he had played in Syracuse, allowing his work to be both vocation and avocation. His club memberships—Rotary, Union League, Masons—became occasions for meetings or conducting informal business. With his father comfortably settled in his post as pastor of the First Methodist Church in Columbus, the younger Peale was no longer concerned about his job security, although he worried about his mother's increasingly fragile health.

An early encounter with criticism, at the very onset of his ministry and from an unexpected source, temporarily postponed the unveiling of his still-congealing message of practical Christianity. It came from within the church, not outside, and thus challenged Peale on territory he customarily felt was secure. Significantly, one of the elders on the Consistory, a man with status and authority in the church, thought that the theme of Peale's inaugural sermon, drawn from John's text about abundant life, was inappropriate.

Peale had prided himself on his ability to work harmoniously with the men on his governing boards. Moreover, he was dismayed that the elder took exception to the implications he drew from what had become his signature text. To the elder, an old-time Calvinist, the sermon conveyed the image of a bootstrap operation and was therefore not only untimely in the midst of a depression, but oblivious to an individual's helplessness in especially hard times. The substantial issues in the ensuing struggle between Peale and the elder were soon lost. Instead, the argument was about lines of authority, which for Peale meant wrestling with his own internal problems about authority figures and appearances of control, as well as with his inability to deal with contentiousness and criticism. Through conversation and letters, he defended the need to encourage people to have self-confidence, but the elder was not persuaded and continued to be a thorn in Peale's side over the next several years.[9] A foe so formidable was, for Peale, a persuasive argument against launching a new approach the man opposed.

The people in the congregation, largely unaware of the dispute, showed their support for the new minister by gradually—painfully slowly, for Peale—filling up some of the rows of empty pews in the church. In the pre-World War II years, Peale had three opportunities every week to address his flock—twice on Sunday and once on Wednesday evening—and he created new, original presentations for each service. Because of the elders' criticism, he generally withheld his message of practical Christianity for the first two years of his pastorate, relying on more conventional sermons, which were nevertheless still creatively titled as "The Eternal Contemporary" and "Stars and Broken Hearts." The one restriction he tried to impose on himself and was least successful in observing dealt with his deep desire to blast the policies of the New Deal. In one of those ironies of history, both Peale's New York ministry and the New Deal were launched at virtually the same time.

He promoted the image of Marble as a "friendly" church, its motto, "Where you are a stranger but once," printed across the Sunday bulletin. Marble had a few old money elite in its congregation, but Peale's hope was to attract more individuals like himself: new to the city, rootless, looking for a church home, people who would not be put off by the Fifth Avenue address and Tiffany windows. An active men's club brought men into the church building, but critics complained that few of them had any other association with the church or its program. Overall, the congregation resembled most other urban churches. It was made up of more women than men, most middle-aged or older. For many of these worshipers, the church was a surrogate home, and the message Peale eventually spoke to them of self-assurance and self-fulfillment was intended as much to combat the disconnectedness of city life as the consequences of the Depression. The new members he attracted to the church were drawn largely from the ranks of the urban middle- and lower-middle-class business and commercial corps, with a sprinkling of professionals at the upper end of the social scale and more clerical workers at the other.

As in most Protestant churches, women predominated. Voters barely more than a decade, the women in Peale's congregation were widows, housewives, some divorcées, and pink collar workers, with political sensitivities just developing. Like many other previously marginal groups, most were doubtless prepolitical in their views of social change, perhaps particularly where their own interests were concerned. Once the enthusiastic supporters of the fight for prohibition, they became quiet participants in Peale's redirected struggle against the New Deal and then part of a silent army of "Cold Warriors." History and tribal loyalties, as well as their minister's ability to recast political issues in terms relevant to their interests, kept them in the column of conservative Republicanism or else silent about their opposition. They listened attentively to sermons illustrated with male heroes and filled with masculine allusions. If they found empowerment in their lives as a result of Peale's sermons, it was generally not in the realm of partisan politics.

For Peale, it was a constant battle to keep his political tendencies in check, enmeshed as they were in his religious outlook. Actively, aggressively political by the end of the decade, he was more circumspect during his first few years in New York, limiting himself to sharp comments in sermons and talks. In his very first year at Marble, Peale took on the Roosevelts—Eleanor and Franklin—publicly and in print. Mrs. Roosevelt had quickly brought the public to confront the Democrats' position on Prohibition: Voicing her Brahmin opinion, she said that Prohibition, by encouraging illegal drinking establishments, replaced the way young people—especially young women—were traditionally introduced to alcoholic beverages, at supervised family dinners and in private clubs. Young women, she observed, would have to learn how to drink socially and responsibly.

Straining to respond, Peale fired back at a Prohibition Day luncheon of the Women's Christian Temperance Union with a rebuke that surely cheered the heart of any populist: "I do not like to publicly criticize a lady," he said,

> especially the next first lady of the land, but in the name of heaven, how could she stand up and say that every girl early in life must find out how much rum she can hold. Her knowledge of the United States does not go west of the Hudson river, and yet there is this statement by this child of the rich who does not know anything about American life. I can't say her husband is much better.[10]

He faulted her for her politics, economics, and sociology; that is, for her views on prohibition, her wealth, and her eastern bias. He would continue throughout his life to see these three issues as a single devil to be wrestled to the ground. For Eleanor's husband he had similar chastisement: He had set a bad example for Sabbath observance by leaving for a fishing vacation on a Sunday and by going in a sailing vessel, an artifact of the affluent.[11]

Essentially a conservative anti-New Dealer, Peale opposed the worst of the Depression demagogues and could on occasion join with liberal social activities, especially when the project was a local one. He signed a published memorandum opposing Franco's government in the forties; and in 1934 he helped create a local committee of the clergy on housing to deal with slum reduction. The latter was an ecumenical commission that worked with city and federal officials to develop proposals for combating the housing problem.[12] When he was attacked for his political conservatism, Peale would point to activities such as these to explain that he was an independent who responded to issues rather than party loyalties. It is possible that there were times when some Democratic candidate who supported a Peale position gained his vote, but it appeared as if Republican conservatism was in his genes.

His political preferences put him well within the Protestant mainstream during the Depression, in harmony with the middle-class constituency he served. For unlike theological priorities, political interests could be tested through the franchise, and as liberal spokesman John Bennett observed, the success of the New Deal was manifest "in spite of the opposition of the majority of members of the Protestant churches."[13] Liberal Protestant leaders of the period were not generally in the churches, but in the seminaries and church agencies, especially the Federal Council of Churches.[14] For most rank and file Protestants, politics and religion had coalesced around the prohibition struggle, now a dead issue, and with Roosevelt cementing Al Smith's 1928 antiprohibition coalition, the Protestant establishment they had taken for granted seemed further endangered. Roosevelt's symbolic appointments of Roman Catholics and Jews to visible public offices—vivid evidence that immigration had actually produced cultural pluralism—encouraged Protestants to focus their status anxieties on the New Deal. It also fostered reaction: some of it demagogic and extremist, as with the resurgence of the Ku Klux Klan; and some of it, like Peale's, an invigorated tribal confidence that the church could be the vehicle for restoring Protestant hegemony. Those few church leaders who flirted with ideas on the extreme left were presumed to be disloyal to their heritage, deserting the majority of Protestants who reinvested their Prohibition passions in the anti-Communist crusade. The New Deal was the seedtime for the harvest of postwar anti-Communism, and for many Protestants, it was a religious effort as surely as Prohibition had been.

These internal polarized struggles of Protestantism, real and imagined, became the special focus of Peale's public platform. His particular targets were those liberal church leaders, privileged by education, birth, and breeding, he believed, as Mrs. Roosevelt was, who chased after New Deal attractions while ignoring a ministry to "ordinary" individuals and neglecting personal religion.

Protestant liberal leadership, to be sure, trod a slippery slope; while mindful, and probably supportive, of the inroads that psychological pastoral care had made in local parishes, they assumed as their own responsibility the difficult task of making moral judgments about the nation and the culture, and acting on them. To people like Peale, their focus seemed as skewed as his did to them. From his perspective, they had abdicated their task of reshaping secular culture around the old Protestant values. At the time, their failure persuaded him that the most effective way to restore the authority of the past was to revive "the Church" and enlarge its membership.

Transformations and Conversions

For the first two years of his New York ministry, the "marvel of Methodism" discovered that Marble Church was more resistant to his considerable abilities than his previous churches had been. Compared with the history-laden Collegiate Church, the others had been ecclesiastical youngsters, and because of their Methodist foundation in lay authority, they had been open and amenable to Peale's direct evangelistic approach. The Collegiate Church, weighted by history and hierarchy, moved slower and was less likely to reflect sudden change. Accustomed to expecting growth, Peale was disappointed that the number of new memberships did not rise more rapidly. At the end of six months, the Collegiate Church *Yearbook* reported that at Marble "there has been a steady, normal increase in the attendance both at the morning and evening services, with an unusual number at the midweek meetings."[15] In the spring of 1933, an "every member canvass" to increase the budget produced forty-one new subscribers and an additional one thousand dollars in income.[16] Despite the limited membership growth, Peale was discouraged to the point of regarding himself a failure.

For his first summer holiday as Marble's minister, he traveled with Ruth to Europe, setting a pattern for overseas vacations that would continue throughout their lives. Such sojourns provided a means for gaining distance from problems, while the details inherent in travel gave him the kind of limited relaxation his restless, energized system could tolerate. The young couple, with their first child expected in the fall, visited Oberammergau, before rendezvousing with Norman's parents in Ireland so that his mother might visit her own father's birthplace. This occasion allowed Ruth to move further inside the tight Peale family cocoon.[17]

During his second year in New York, the pace of family life picked up. The first of three Peale children, Margaret Ann, was born on November 17, 1933. She would be joined three years later by brother John Stafford, born September 2, 1936, and six years after that by sister Elizabeth Ruth, born July 22, 1942. Almost instinctively family-oriented, Peale tried to recreate with his own children the family bonding that had held him to his parents, though for most of the children's growing-up years he was either on the road or involved in church activities, leaving the primary care to his wife.[18] Even with new baby Margaret Ann, Ruth remained active in the church, dutifully following the requisite role of the pastor's wife, serving on church committees and putting in her time as the secretary of one of the women's groups.

Peale concentrated on his preaching, preparing sermons in his customary way for "dealing with the practical problems of hard-pressed men and women in these troublesome times." Church services were "uniformly well attended," with 131 new members joining the congregation during the calendar year of 1933. That same year Peale announced that he was devoting more time to "personal consultations regarding spiritual matters," the tentative beginning of the therapeutic counseling program.[19] At a time when Baptists and Methodists were losing members and Presbyterians just holding their own, Marble's limited growth bucked a trend. But Peale was still discouraged, still preaching to a sea of empty pews.

Then in 1934 his ministry changed course. From his point of view, the dramatic turning point occurred the following summer when he and Ruth were once again traveling overseas, this time in England. They stayed at Keswick in the Lake District, home of the famed summer evangelical conferences of the Church of England. Here, as Peale recalled the event on many occasions, he experienced another spiritual transformation, a "conversion." According to Peale, he and Ruth walked together in the garden outside their hotel while he unfolded his anxieties and disappointments about the lack of progress he had achieved at the church. As on a similar outdoor excursion with his father when he was a boy, he was instructed to sit down and pray for renewed faith and trust in God, to "get converted" once again. This time it was Ruth, not his father, who spoke to him with firmness and conviction, he recalled, she sitting beside him, holding his hand while he prayed aloud, confessing his weaknesses and "surrendering" himself to the Lord. Immediately, he believed, his prayer was answered, and he felt "warm all over," his energy and his commitment to the church restored. Much like his earlier conversion experience when he had reported it, this, too, he named a life-changing event, one that restored his will to tackle the difficulties at Marble Church.[20]

There was likely a psychological overlay to this event which fixed its impact in his memory. Its fuller implication may be hidden in a 1955 article Peale wrote for *The Reader's Digest*, a piece that reflected his tendency to fill his writings with strong autobiographical references. In the article, describing the problems of a young minister whose insecurities had caused him to consider leaving the pastorate, Peale supplied this lengthy revelation:

> A brilliant young country preacher, he had been discovered by a summer visitor and summoned to the pulpit of an important church in a large city. Sensitively conscious of his youthfulness for such a position, he worked under heavy pressure, largely self-imposed. He came to us [at the Marble clinic] sleepless, unable to eat and extremely nervous.
>
> Pacing the floor, wringing his hands, he blamed himself for accepting the new place. "I'm just not up to it," he repeated desperately. "I can't take the strain—producing highly intellectual sermons for a city audience. Other ministers expect me to fail—they want me to fail."

> I let him pour out the mass of discontent and pathetic self-doubt without interruption. Then we talked. In great part, his inferiority feelings owed their rise to an atmosphere of insecurity and anxiety in his so-called "Christian" home as a boy. While the Bible and prayer were constantly in evidence, Christian virtues of faith, trust and peace were sadly lacking.
>
> We cleared away much mental debris . . . [and] this understanding in itself had an important curative effect. . . . Therapy continued over some weeks. . . .
>
> Then we made positive suggestions. The moment when the young minister felt most insecure was just before the sermon, while he sat in full view of the congregation, waiting to mount the pulpit. We instructed him to pray at this moment, not for himself, but for the people in his audience, thinking only of how he might help them individually to find power, peace and happiness through leading useful lives.
>
> We urged him to become indifferent as to how his sermons were regarded and to practice having a better opinion of his ministerial colleagues toward a fellow clergyman. At any rate, his task was not to please them, but to help people. His only thought in composing a sermon, therefore, was to be how it would help people solve the problems of their lives. The simpler, more practical and more direct he made his addresses, the better.

The anonymous minister heeded the advice and began preaching sermons that were "down-to-earth, as they had been before he came to the city," and as a result, he "not only remained in his church, but in the course of time has come to be the wise and valued counselor of hundreds."[21]

With the young minister serving as Peale's alter ego in a kind of therapeutic confessional, this account may reveal not only how Peale perceived his Marble ministry in the pre-Keswick years but how he handled other anxieties. The autobiographical references provide rare clues to some of Peale's inner fears: the expectations for a big city church, the fear of clerical criticism, the fear of failure, the insecurity of sitting alone in front of the congregation, the Christian home that was less than perfect. Just when Peale himself gained a perspective on parallel tensions in his own life is not precisely documented, yet the evidence speaks to some important changes that occurred in his early New York years that resembled those of the young clergyman. It was at this point in his ministry Peale made his one and only return visit to the Boston University School of Theology, feeling totally rejected by his audience. His local ministry was disappointing, and perhaps, he wondered, even a bad choice. He felt intimidated on the church platform, criticized by his own board.

And, separated from his natal family while steadily hearing his mother rehearse her illnesses, Peale may have been encouraged to re-examine his childhood memories. In 1940, his candid associate, the psychiatrist Smiley Blanton, advised him to consider the roots of his inferiority complex, hinting that the two had spoken of the problem earlier. A reasonably honest appraisal would have convinced him that his childhood had not been unclouded, that in addition to the financial insecurity, the status anxiety, and the steady moving, there was also the hectic, sometimes harried, schedule of both parents. What he lightly referred to as "the whole world" regularly trooping through the family dining room may have produced in the sensitive "special" child tension and dread rather than color and excitement.

Whatever the ingredients that combined to refocus his appeal, they "worked," as thousands later claimed his healing message "worked." His ministry gained new dynamism. Seeming at last to accept the fact that he would not receive universal acclaim for his unusual message—though still sensitive to criticism—Peale finally gave in to his urge to preach thoroughgoing practical Christianity. And the church and his waistline grew steadily. At the end of two years, he had added 237 members, with more impressive gains coming in the next several years. Some observers pointed out that change was attributable to New York's gradual climb from the trough of the Depression. Equally, at Marble, it was due to Peale's new approach.

Creating a Style

The first change apparent to the congregation was essentially a symbolic statement about the centrality of preaching in the worship service. At an earlier time Marble had adopted an open chancel arrangement, which located the minister's pulpit chair in the center of the platform, flanked by chairs for assistants, the clergy, and the congregation, so that all would gain an unobstructed view of each other. A small lectern stood off to one side from which the minister could preach or make announcements. It was an exposed position, such as the one that frightened the young minister in Peale's *Reader's Digest* story. Peale abandoned the lectern and spoke directly to his audience, as evangelists before him had done, without a barrier between them, bound to each other in a kind of metaphysical embrace. He carried no notes and had nothing to lean on. In addition, partly to guard against presermon jitters and partly to delay the emotional engagement of the congregation until the sermon, Peale cultivated a blank, solemn, sometimes sullen, appearance during the opening parts of the service, looking without seeing, as if his audience was not there.

Speaking three times a week without notes required a change in the preparation and presentation of sermons. Peale began a technique he called "picturizing," adapted from the mental imagery of New Thought, which he would later refine and describe as "imaging." In time he developed a formula for creating sermons consisting of three points, illustrated by anecdotes, and held together by a theme adapted from a biblical text. In crafting the sermon, he wrote a full outline and "picturized" it, a process he likened to taking a photographic snapshot of it with himself preaching it.[22] Imaging a desired goal was a form of mental suggestion common to the religious science experience. In time it gained acceptance across a broad spectrum of American culture, invoked as much by physicians treating cancer patients taught to imagine themselves winning a deadly combat against cancer cells as by athletic coaches hoping to motivate students to win their races.

Peale called himself an evangelist and compared his preaching style, with its implicit call for conversion, to what the old-time revivalists called "drawing in the net." His public style was animated with visual attraction without being flamboyant. To sell religion—and he admitted it was selling—demanded such personal commitment and involvement with the subject that notes could only be intrusive. Think of it analogously, he suggested, to selling a car:

> You would not say, well this is a pretty good car, and let me check my manual and I can tell you more about it. No, you would look him in the eye and say, "Look friend, this is it, and Bill Smith got one, and this one is just like it and it is the best."[23]

Preaching was a way of advertising the gospel.

The ideological underpinning for his message was quite simple. Its major derivation was the familiar text from the gospel of John about abundance. As a young Methodist, Peale had first understood the text in terms of the spiritual abundance attendant on the conversion experience, which opened up a future life that was full and satisfying. In the thirties, he repeated the theme, only with harmonies and a descant. The resonance for him was best when it involved what Josiah Royce had called "the rediscovery of the inner life," a spiritual center where Peale believed one made contact with the energizing Presence. Some twenty years later, when Clifford Peale analyzed the elements in his son's composition, terming it "a new Christian emphasis," he described the product as "a composite of Science of Mind [New Thought], metaphysics, Christian Science, medical and psychological practice, Baptist Evangelism, Methodist witnessing, and solid Dutch Reformed Calvinism."[24] Even his father had come to realize that with these variations his son had created a wholly new composition, having parts drawn from the religious mainstream, some from the margin, and others from the secular culture.

In fact, had Peale's inclination been toward a form of systematic theology, there were really no theological options open to him. He disliked fundamentalism and was even more uncomfortable with Neo-Orthodoxy than he was with social Christianity. In any case, it appeared to him that liberal theology had not only lost its spark, it was losing its support, as was vividly demonstrated at the Methodist General Conference in 1936 when, retrenching on social programs, the delegates reconstituted the Board of Education with the conservative Bishop Adna Leonard, Peale's patron, as its head.[25] A practical religion attuned to the inner life, Peale believed, was a better solution to the needs of modern Americans.

In 1934, under the sponsorship of the Radio Department of the Federal Council of Churches, Peale took his message to a national weekly audience, his program, "The Art of Living" carried by affiliates of the National Broadcasting Company from May through September. It provided him with an enlarged identity as a human interest preacher of practical Christianity. When he invited letters from his listeners, his office was quickly swamped with request mail from people asking for advice or prayer and from others simply offering support.[26] Peale was an especially strong radio speaker, more effective than he ever became on television, and radio served him well in creating a vast national constituency around his approach to practical Christianity. His sponsorship by the Federal Council of Churches revealed that there were areas in which his presentations enjoyed the support of mainstream agencies. In the early forties the Reformed Church also indicated its support for his preaching skill by inviting him to teach a course in homiletics at the denominational seminary in New Brunswick, New Jersey, something Peale did for several years. Whatever it was that triggered the change, Peale's ministry began to surge on many fronts after 1934.

The Message of Practical Christianity

Peale's sermons during the decade were a mixture of religion with politics, of celebrations in the church calendar with reflections on the passing cultural scene. Intended only for a single presentation, they were important indicators of the immediate issues that claimed his attention.

His relatively more considered reflections on religious values at the time appeared in two inspirational books he wrote during the thirties: *The Art of Living*, published in 1937, and *You Can Win*, which followed the next year. With Peale, form and substance were one, and thus his writing was often as turgid as his preaching. The two modest books revealed that Peale had become immersed in the literature, and probably the beliefs, of metaphysical spirituality, or religious science.

As examples of the inspirational genre—metaphysical, redolent with affirmative language of New Thought—the books are unexceptional. As evidence for the religious values he hoped to convey to a national reading audience in these, his premier presentations, they are historically significant. A well-known mainstream minister, Peale revealed his sympathy for a religious science position some distance from orthodox Christianity.

By the time his first book appeared, there was an already existing model for New Thought inspirational writing, the standard set by Ralph Waldo Trine with his *In Tune with the Infinite*, although works by Orison Swett Marsden and Henry Wood followed close behind. Peale wrote in a similar vein, drawing on New Thought imagery about God, Jesus, the sources of faith, to develop his message of hope and optimism. It was a characteristic of writers of this school to employ familiar religious terminology with nuanced, idiosyncratic meanings, so that old associations with such basic concepts as God or Jesus were often undefined. In Peale's case, readers would have had a difficult time deciding whether his language was a reflection of his metaphysical preferences or an aspect of his unreconstructed Methodism.

New Thought writers used a variety of terms to describe God. God was Mind, Omnipresent Spirit, All Supply, the Higher Consciousness, the Christ Within. The Divine Presence resided in organic unity with Nature, and was made manifest in the human spirit. In this context God was then an immanent force, pervading the universe, reflected in human higher consciousness, and potentiating in power, energy, and growth. Further, this God was loving and good, devoid of evil, and ever ready to bestow health, peace, and well-being on those who, in Trine's words, got *In Tune with the Infinite.* Additionally, time and existence were cyclical, not linear, the human essence constantly in process, enjoying a pre-life and an afterlife that assured personal immortality. Finally, the New Thought-metaphysical understanding of Jesus resisted the traditional Christian concept of incarnation, often describing Jesus as the "way-shower pointing out the path of Christ-consciousness within."[27]

Peale's books talked about how to make living an art, and more specifically how to locate and actualize the power to make that possible. Power was an important theme in the literature of New Thought from at least the 1890s on. There was note of irony involved in using the theme, given that the primary audience for religious science was middle-aged, middle-class women. Power therefore tended to be defined within a narrow personal context, primarily as the ability to control one's own life, health, and destiny. But power could also be described in robust terms of financial success. If men appreciated the potential of religious science literature for political and economic application, it is likely that women understood it in terms more available to them, as personal empowerment, a kind of personal, pietistic feminism. In Peale's two slim volumes, God was readily identified as the source of power, with Jesus "the perfect channel" through which the power flowed, the perfect model for others. Power was active and dynamic, and therefore the opposite of that kind of Christianity that depended on tradition and what Peale referred to as inherited "passively held" beliefs.[28]

Explaining how to get in tune with the source of this power required more elaborate imagery. Peale explained that "You bring your spirit into harmony with the Spirit of God. . . . You say with the faith of a little child, 'Lord, I bring my human spirit to you and I ask you to fill me with your power.'" And then, if you are serious and really in tune, "the miracle happens." Peale employed manifold cultural imagery to describe how the individual could be filled with the power of God: It was like an orchestra filling a room when a radio is turned on, or when a dynamo is attached to the source of power, or what happens when sluice gates are opened, or a light bulb is screwed in, or a meandering stream joins a mighty torrent, or a stray speck of dust is removed to empower a malfunctioning elevator.[29] The message was that the believer was to be attached and in tune with the source of power, not detached and isolated from it.

The primal, almost libidinous, implication of his location of the source of power was more vivid in his allusions to power channels. People, he contended, were not "intended to be merely receptacles but channels of energy." Those who opened themselves to the "spiritual Christ" would clear internal channels to allow the divine power to flow through. As channels, individuals needed to stay connected to the source of power: "When we detach ourselves from this flow of power, we become isolated units and gradually become inwardly barren, with the result that all manner of infirmity and disease develop within us."[30] Many women could be especially vulnerable to feelings of inner psychic barrenness when their familiar lives suddenly became isolated and detached in middle age.

But connected to the source of power, unlimited potential awaited. Affirmation, another of the major ingredients in the New Thought corpus, was also evident in these two books. Citing the biblical text "I can do all things through Christ who strengthens me," Peale applied it with a literalness he would use repeatedly. One needed to believe the text, and believe in it fully, and then affirm it through prayer.

New Thought and religious science literature generally, like the tradition itself, frequently cited the compatibility between its spiritual claims and general scientific principles: Empirical evidence was said to support its testimony. It was within that context Peale wrote about the law of averages, which for him meant that most setbacks were only temporary, that "the good never loses." He explained reassuringly in *The Art of Living* that "Life averages well, which means that most of the troubles we are worrying about now will never happen." The overly examined life would inevitably discover problems; therefore, "The best way to live is just to live."

The books were not, however, singular reassertions of New Thought concepts. Intertwined with the optimism and affirmation were elements of Methodism, the Oxford Group, and even hints of Calvinism. Peale spoke of sin and struggle and the need to surrender one's life to God. In *You Can Win* he wrote, "Simply stated it means: A man has no power in himself, he is weak; his will is weak; he wants to gain a victory but does not have the strength," which can come from God.[31] Like the program Bill Wilson was then assembling for Alcoholics Anonymous, the prescriptions contained in the books were meant to show how people could gain control over their lives. It was part motivational theory, part mystical pietism, and part primal myth. It suggested boundless, unqualified power at a time when the cultural scene still offered soup kitchens and Hoovervilles.

A Psychotherapeutic Healing Clinic

The local environment offered abundant evidence of the difficulty people found gaining control of their lives. From the very outset of his New York ministry Peale reserved eight to ten hours weekly for individual sessions with persons seeking his help. As the list of potential clients grew longer, Peale approached his own physician, Dr. Clarence Lieb, in 1934, for help in finding a professionally trained therapist to assist him in the work. He was hoping to implement locally a relatively new concept he knew was being tried elsewhere.

In 1934, when Professor John McNeill observed, "We are evidently at the opening of a new era in the history of the cure of souls," he was reflecting a sentiment that prevailed in many quarters.[32] Forms of psychotherapeutic pastoral care were finding a niche in mainstream Protestant churches as local pastors learned more about developments in psychology and their relation to certain religious concepts. From the turn of the century, and with public interest aroused by the Emmanuel Movement, seminaries and urban churches had shown a willingness to experiment with the application of psychological theories to problems in pastoral care. Peale's famous New York neighbor at Riverside Church, the Reverend Harry Emerson Fosdick, became a notable convert to the importance of counseling, observing that "My preaching at its best has itself been personal counseling on a group scale."[33]

Pioneers in pastoral psychotherapy, such as the Reverend Anton Boisen and Dr. Richard Cabot, tried to shape it into a unique, integrative discipline, with its own theoretical supports and clinical training programs. The new field of psychosomatic medicine was a product of this fertile environment, finally taking its place in the thirties as a challenging subspecialty in medical practice and research. One of the early leading figures in the field was Dr. Helen Flanders Dunbar, confidante of Anton Boisen's and once a friend of the Emmanuel Movement's Elwood Worcester.[34] Dunbar's appreciation of a holistic approach to healing was evident in her important study, *The Emotions and Bodily Changes*, and was continued through her editing of the new *Journal of Psychosomatic Medicine.* Dunbar's appointment as the first director of the Council for the Clinical Training of Theological Students was additional evidence for the compatibility of psychosomatic medicine and pastoral care. Still not mainstream in either religion or medicine, forms of holistic healing that considered the mind an integral component in the therapeutic process were gaining selective, but well-placed acceptance.

Depression era problems seemed to create a need for a special ministry for the cure of souls. With such considerations in mind, Peale asked Dr. Lieb for professional assistance, a request he in turn referred to Dr. Iago Galdston of the New York Academy of Medicine, who then passed it on to the Freudian-trained psychiatrist Dr. Smiley Blanton.

First over lunches at the Harvard Club and then in sessions in Peale's office, Dr. Blanton and Peale devised a treatment method that was more professional than Peale alone had been able to offer. Once they put the new format in place, Peale continued to see patients first, but he consulted about the cases with Blanton, and, depending on the nature of the problem, he counseled the patient himself based on treatment worked out with Blanton, or sent the person directly to Blanton. It duplicated the model Elwood Worcester had developed at the Emmanuel Clinic years earlier. By 1937 Peale and Blanton decided to set up special offices, which they referred to as the Marble Collegiate Church Clinic, in the basement of the church, where they continued to see people—largely from the congregation—on a regular basis.

Peale had stipulated that he wanted a "Christian" counselor for his associate in this project, and with Blanton he came as close to that requirement as he was apt to get within the field of psychiatry. Describing himself as a "Tennessee Methodist hillbilly," Blanton was born in Unionville, Tennessee, in 1882, eventually obtaining his undergraduate education at Vanderbilt and his medical degree from Cornell. In 1923 he was awarded a Diploma in Psychological Medicine from the Royal College of Physicians and Surgeons in London. He had served in World War I before settling in New York in 1931, where he devoted most of his professional time to teaching. He continued teaching even after joining Marble's staff—partly to pay off debts he had accumulated—first at Cornell, from 1933 to 1938, and then at Vanderbilt Medical School from 1943 until his retirement.[35] In 1929 he spent time in Vienna being analyzed by Freud. He returned for more analysis with the famed architect of the psyche during the summers of 1935, 1936, and 1937.

As much a Democrat as Peale was a Republican, Blanton was seen as feisty, earthy, shrewd, and neither excessively orthodox nor heterodox in his interpretation of Freudianism. In fact, he seemed most comfortable with neo-Freudianism. His training with Freud, however, had made him more negative than Peale in his assessment of human personality, willing to see primal drives as a constant threat to the ego. Possessed of a charismatic personality, he reportedly achieved almost instant transference with patients, some of whom later found it difficult to break away and individuate.[36]

In these early years of the clinic, the minister and the psychiatrist were not so much blazing new trials as adapting the model in use since the heyday of Worcester's Emmanuel Movement. Pastoral care as it surfaced in mainstream churches dealt with what Worcester had called "functional illnesses," as opposed to organic or somatic problems. Records describing the patient population after fifteen years of operation indicated that the Marble Clinic treated essentially the same problems Worcester had addressed—anxiety, depression, morbid fears, alcoholism, "inadequate social relationships," and unhappy marital and family situations.[37] As at Emmanuel, individuals with more serious problems, usually of an organic nature, were referred elsewhere.

In its theoretical perspective, the clinic was eclectic. Over time, it developed a theoretical base that was Jungian, with strong evidence of neo- and post-Freudianism. Blanton brought to it a neo-Freudian frame of reference. Peale, while claiming to draw from not just a single source for his psychological information, was more familiar with Carl Jung than with Freud or other theorists. There was an obvious affinity between Peale's view of the mind as a divine link with the cosmos and Jung's inward guest for the "original experience." The minister's claim that individuals not connected to the divine Presence were incomplete and probably unhealthy, had its parallel expression in the psychiatrist's contention that people experienced "a never-failing homesickness for the *Pleroma*, the 'fullness of Being.'" Jung's inward questing led him to develop his theory of universal archetypes, and also to his 1916 work *The Seven Sermons to the Dead*, with its heavy indebtedness to esoteric views including that special quality of knowing, or Gnosis.[38] Many practitioners in the field of pastoral care found Jung a relevant source, and it would hold as true at Marble's clinic as at others. Peale himself tended to abjure theory and systematics in any field, so he could more likely be described in his counseling work as atheoretical, his views shaped more by his interactions with Blanton than by any other source.

By 1940 Peale had begun to withdraw from the daily operations of the clinic, his crowded schedule booked with speaking engagements months, even years, in advance. As the caseload increased, Blanton, appointed director of the clinic, was joined by other ministers from the church staff and by part-time psychologists and psychiatrists. For years their work was subsidized by Marble Church, but as the client base grew larger and extended deep into the community, it was necessary to charge small fees. With operating costs continuing to rise, Peale assumed the role of fund raiser, diverting a significant portion of his lecture honoraria to the clinic.

Transformed from its basement image, the clinic continued to have a number of different lives. After Peale's withdrawal in 1940 and the addition of more professionally trained staff, the clinic, though still sustained by the church, secured a separate identity as a "religio-psychiatric" agency. The new professionals introduced a broader theoretical base, while concentrating largely on such neo-Freudians as Jung, Alfred Adler, and Erich Fromm and later the post-Freudians Harry Stack Sullivan, Abraham Maslow, Rollo May, and especially Carl Rogers, whose client-centered approach became almost synonymous with pastoral counseling. Reliance on this body of neo- and post-Freudian theory exposed the clinic, in its various guises, to the criticism of the psychologically orthodox, largely Freudians, who termed it an approach of "Sunday sermons. . . . The positive is promoted so as to drive out the negative. One strives to be cheery because it is a cheerless world."[39] From this critical perspective on the political left it seemed as if bourgeois religion and bourgeois therapy discovered their mutual attraction.

The clinic entered another phase in 1951 when it became known as the American Foundation for Religion and Psychiatry, organized as an interdenominational, nonsectarian agency with its own board of directors and a new home on Park Avenue. The following year the State Board of Welfare licensed it and approved it as a psychiatric clinic, with a proviso for screening patients to ensure that those needing such care would see a psychiatrist.[40] Its annual patient load climbed steadily, from about 3,000 in 1952 to 25,000 in 1962. Aware that over 40 percent of persons seeking help turned first to their ministers and rabbis, the Foundation began a clinical training program in 1952 for the clergy and religious educators, which in 1968 was separately incorporated as the Blanton-Peale Graduate Training Institute providing an intensive three-year curriculum.

The Foundation relocated again in 1955, moving into new and larger quarters adjacent to Marble Church on 29th Street. Funding continued to be a problem; subsidies from Marble Church and patient fees could not keep up with an expanding program. Peale's contributions from his speaking honoraria and charitable donations from his wealthy friends helped make up the deficit. Peale's friend J. H. Kresge, through the Kresge Foundation, helped with the expense of the new facility, and Alfred P. Sloan underwrote efforts to expand the clinic's work to include research. Other major contributors—Josephine Bay, Arthur Rodenbeck, Mrs. Kerwin Fulton, Mrs. Albert Lasker, and especially Clement Stone—were directed to the clinic by Peale.[41] Stone became chairman of the board of directors in 1962 and was a generous contributor for a number of years, his resources deriving from a multimillion dollar insurance business he had built. A high school dropout who fit the idealized model of Peale's self-made man, and who attributed his success to a positive mental attitude, or PMA, Stone described his strategies in such works as *Success through a Positive Mental Attitude* and *The Success System That Never Fails*, as well as in his monthly magazine, *Success*.

The Foundation had its next, and most recent, incarnation in 1972 when it merged with the Academy of Religion and Mental Health, which had been started in 1954 by the Reverend George Christian Anderson in collaboration with Paul Tillich and others, to become the new Institutes of Religion and Health (IRH). The home organization expanded to include over thirty full-time professionals, working in either the New York center or one of the three suburban branches. The IRH published the *Journal of Religion and Health* and achieved national recognition as a distinguished mental health organization with twenty-five affiliated counseling centers. The Clinical Affiliates Program had been partially funded by a large grant from the National Institutes of Mental Health, an acknowledgment of the perceived effectiveness of the IRH. From its beginnings in the basement of Marble Church, the IRH had become the largest psychotherapeutic pastoral counseling, treatment, and training organization in the world.

Peale's relationship to the IRH after he withdrew as an active participant in the early 1940s became that of fund raiser and chief executive officer. He was called in when contending personalities required mediation and again when contentious issues—sometimes raised by church members wary of the Institutes' liberal policies on sex and religion—needed resolution. Peale understood that the clinic's approach differed substantially from his own, but he did not attempt to intrude on its work. When, in fact, his involvement seemed more a liability, he tried to downplay his role. Once when Peale was asked by Blanton why he had agreed to let Dale Carnegie offer one of his oversimplified courses on public speaking at either the clinic or the church, Peale offered a revealing explanation: He said, "one of my failings is a kind of easygoing, all-things-to-all-men attitude. In other words, if anybody suggests anything I always try to do everything to please everybody. This is a poor attitude and has gotten me into hot water more than once."[42] Peale deferred to the judgment of Blanton and the clinic's staff and withdrew the offer to Carnegie. Another incident occurred when the clinic planned to submit a grant proposal to the Rockefeller Foundation, a member of whose board was Henry Pitney Van Dusen, President of Union Seminary. Convinced that although Van Dusen was a fair man, the Peale name would raise a red flag for him and defeat the proposal, Peale asked to have his name removed. Again, in the wake of the controversy over *The Power of Positive Thinking*, Peale offered to remove himself from the scene lest his popular image diminish the effectiveness of the IRH. He remained as president of the board, a largely advisory role, with the magnetic ability to draw in money for the work.

A Missionary to American Businessmen

Peale's contention that his public message was as much for himself as for his audience was never better evidenced than in the last half of the decade, when his breathless, quickened pace supported his claim that for the spiritually-energized there were opportunities available, even in hard times. His schedule was jammed, people waited months for an appointment to see him; and he developed complicated arrangements with his secretaries to convey messages between them and himself. If such potential existed during the Depression, what might a restored future hold?

In addition to preaching at Marble, working in the clinic, and writing books, speaking engagements filled the remaining spaces in his calendar. Peale accepted summer preaching appointments at the religious meeting ground in Ocean Grove and at Elberon, both in New Jersey. For several years he was part of a team ministry with John Sutherland Bonnell, of Fifth Avenue Presbyterian Church in Manhattan, conducting weekday services at local hotels. But his most lasting speaking commitment was to the business community, ministering to its members in their territory rather than his.

At a Harvard Club luncheon in the late thirties, "one of the determinative moments in my life," Peale later recalled, he learned from some of the "rising young New York executives" that although they were interested in religion and were nominal church members, few actually went to church.[43] He received the news, he said, as a revelation, as a call to become the "missionary to American business."[44] Almost immediately, he signed on with the Harold Peat Lecture Bureau, run by a member of his church, and was soon logging cross-country trips to address many and diverse segments of the business community.

Between the speaker and these audiences there was mutual regard, almost from the first. A master at self-deprecating humor, Peale diminished his own role as "just a preacher" while bolstering their role as important contributors to the health of the nation. He had a limited number of talks that he took on the road, though all were motivational, inspirational, and humorous, with the frequency of repetition allowing him to polish them to high standards of public performance. His powerful talks, self-effacing style, and general good humor made him a welcome guest with salespeople in particular, but also with bankers, railroad men, power companies, and auto dealers. These contacts became links in his growing national network: Some would become *Guideposts* readers; others wholesale *Guideposts* subscribers; and still others generous contributors to the various Peale projects.

As important as their reception of him was Peale's reaction to the business community. He had a particular sensitivity to what he regarded as a "sissy" image of religion and ministers, and he welcomed the opportunity to interact with anyone he could describe as a "man's man." Intimidated and unappreciated at meetings with clerical peers and academics, Peale was free and expansive at gatherings with businessmen. That he went back so regularly and so frequently was a measure of the pleasure he derived from the interactions, which he referred to as "fun."[45] He felt accepted and supported by these secular audiences as he had not, for example, in his talk at the Boston University School of Theology.

Out-of-town speaking engagements began to claim more of his time. Throughout the decade, Marble's Wednesday evening service made it necessary for him to plan his trips around that commitment, but when the service was discontinued in the early forties, he was able to stay on the road continuously from Monday to Thursday several weeks out of the month. Over the years his work in the secular community consumed an ever-larger proportion of his time. He spoke to business groups and civic organizations and frequently to churches in the area, earning fees that allowed him to meet his personal expenses as well as those of his growing list of projects.

So large a time commitment—probably close to a third of his time by the end of the war—obviously reduced Peale's availability for other activities.[46] The major responsibility for rearing his children clearly fell to Ruth, and secretaries and staff managed much of the daily running of Marble Church. But it meant that in a schedule so tightly packed there was virtually no free, unbudgeted time. People wanting to see him often waited months; social occasions were necessarily work-related events; and volunteer activity, such as committee jobs, was extremely limited. To a friend who complained about his lack of involvement with the Anti-Saloon League, he explained that "every organization with which I have been connected would entirely sympathize with you in my lack of value to the work of the society."[47] Clearly the architect of his own life, the designer of his own crusade, Peale reserved little time in his schedule for work in the larger associational community. After 1936, however, despite his whirlwind pace, he managed to create time for political activities, a highly significant indicator of his priorities. When the demands of his radio program and the needs of his young children kept him stateside during the summer, the family vacationed close to New York City so Peale could get to the broadcasting studio on Saturday evening to record his message. He confessed to feeling beleaguered by the pace of things, but he did not let up.

Instead he took on more. There was the constant lure of politics. And presumably for reasons of income, he continued as summer minister at Elberon Memorial Church in New Jersey between 1937 and 1943, a commitment that kept him on trains on weekends as his business trips did during the week. In 1941 he went to Hollywood as technical adviser on the film *One Foot in Heaven* and made a start on writing a monthly newspaper column called "The American Way." Also in the early forties Peale accepted a continuing invitation to preach at Ocean Grove, New Jersey, sometimes speaking ten times in eight days to a fundamentalist evangelical audience he knew held reservations about his message, though he admitted to being sufficiently fluent in theological terms to "speak their language."[48]

In the early war years, thoroughly converted to a practical Christianity based on the positive thinking approach of mental science, Peale began conducting joint midweek services with his New York colleague, the Presbyterian clergyman John Sutherland Bonnell. Meeting at Town Hall and at various hotels around the city, they patterned their approach on that of the extremely popular New Thought preacher Emmet Fox. Their hope, said Peale, was to "out-fox Fox."[49] The services were atypically spare by the standards of traditional Protestant worship, consisting of a little organ music, a period of silence, and then a twenty-five-minute inspirational talk. They discontinued the experiment in 1945 despite the fact that they were drawing audiences of 1,200 to 1,400; random surveys had shown that the congregation was made up mostly of women, and women who were already church members. It was a predictable outcome, and the only thing unusual about it was that the ministers seemed surprised by it.

The war brought economic stability to the nation, but Peale and Marble Church had started to recover from their various crises several years earlier. The mild success of his first book, *The Art of Living*, which sold over 20,000 copies in its initial offering, encouraged his publishers to request the second, and then in 1940 Peale collaborated with Smiley Blanton on still another book, *Faith Is the Answer*. The empty pews at the church were being filled twice each Sunday, and new members began to join in record numbers. In 1941, 203 new members joined, the largest annual addition in the history of the church. With his host of other activities, Peale seemed a veritable one-man industry.

"I Preach for Myself"

But as in all lives, there were shadows. Peale's greatest personal tragedy during the period was the death of his mother in 1939, the first major loss in his life. Mother and son had been extremely close, and he was profoundly moved by her death. Immediately after she died, he had the first of many psychic experiences that involved her. Going instinctively to the church on hearing that she had died, he went to his office and placed his hand on a Bible she had given him. At that moment, he said he felt two hands resting on his head, and "knew" it was his mother's presence. Several years later, after buying at auction a pair of expensive hurricane lamps for Ruth, he heard his mother's voice, reassuring him about the gift and her affection for Ruth.[50] He reported to a friend that the bond with his mother was so strong that for him she had "never really died."

There were also enough besetting problems at Marble Church that in 1943 he considered accepting the attractive offer of a prestigious Methodist church in Detroit. Structurally overextended, the Collegiate Church corporation had fallen into serious debt, and Peale worried about its ability to keep its financial commitments, not to him but to Marble Church. Once news reached the Collegiate Consistory that Peale was considering another offer, it reorganized its financial priorities, reaffirming its continuing support for the work at Marble.[51] The sharp and persistent complaint of a member of Marble Church, which she carried through several organizational levels of the Reformed Church, had provided still another incentive to think about leaving the church. The criticism began because Peale had replaced a paid quartet with a paid chorus at Sunday services, but it soon escalated to include charges that he was frequently away, that he failed to conform to the standards of the Reformed Church liturgy and the theology of the Heidelberg Cathecism, and that he often commented that "this is a good Methodist church." The complainant's dogged persistence in having the charges aired before various local and regional denominational bodies caused the sensitive Peale many sleepless nights. His achievements at the church should have convinced Peale, however, the accusations were predestined to fail.[52]

Despite the visible signs of success, Peale's insecurities continued to haunt him. Working closely with Smiley Blanton, then writing a book with him, encouraged him to be more open with an associate than was customary for him. Blanton, sensing his receptiveness, offered Peale some advice, which Blanton prefaced by saying it was intended seriously. Peale's psychiatrist friend advised him not to "allow this inferiority feeling that you have to influence you so much." Then he attempted to reassure him that his special talents in writing and speaking enjoyed wide support, so that "you need have no feelings of doubt about your ability in these fields."[53] Undoubtedly, many of the women who bought Peale's books and came to hear him speak could identify with his feelings of inadequacy. He said repeatedly that the primary audience for his message was himself. He might draw lightning from the clouds and be charged with cosmic energy for the work of practical Christianity, but it was an energy that needed steady recharging.

Notes

1. NVP to "Dear Folks," Feb. 29, 1932; NVP Ms. Coll., Box 186.
2. NVP to "Dear Folks," Mar. 12, 1932; NVP Ms. Coll., Box 186.
3. NVP to "Dear Folks," Apr. 11, 1932; NVP Ms. Coll., Box 186.
4. Corporate wealth can rarely be determined through readily accessible sources. Peale was told in 1932 that the value of the Collegiate properties was between $15 and $20 million, but by 1943 the Collegiate system had incurred such a large debt that Peale was considering accepting a call to another church. NVP to "Dear Folks," Apr. 21, 1932; NVP Ms. Coll.

In this series of letters to his parents, Peale also revealed that his attachments to his family continued as strong as ever. His younger brother Leonard—who his father said also suffered from an inferiority complex—was living in Syracuse with Ruth and Norman while he attended Syracuse University, and Norman worried about leaving him during his trips to New York and Los Angeles. Peale regularly concluded his letters with endearing expressions such as, "I love you both with all of my heart."

5. NVP to Bishop Adna Wright Leonard, July 15, 1932; NVP Ms. Coll.
6. The clergy assistants at Marble, whose responsibility it was actually to line up new members, made regular reports to Peale about the previous affiliations of members. The numbers changed from year to year, but it is clear that the great majority came from traditional Protestant backgrounds. Peale frequently made a point of noting the significant minority who came from Roman Catholic backgrounds.
7. *The Reformed Church Messenger* complained that the Collegiate Consistory—the lay governing board—so regularly went outside for a minister that it "seemed at times as if there were an unwritten law to the effect that Reformed Church ministers will not be seriously considered." The *Christian Intelligencer* of October 5, 1932 disputed the claims of *The Messenger* by pointing to some Collegiate Church ministers with backgrounds in the Reformed Church. But there was a strong tendency to look outside the Reformed Church when searching for a "name" preacher.
8. Van Varner, *In Celebration of Norman Vincent Peale's Fortieth Year as Minister of Marble Collegiate Church, New York City*, pamphlet, New York City, 1972.
9. Arthur Gordon, *Norman Vincent Peale: Minister to Millions* (Englewood Cliffs, NJ: Prentice-Hall, 1958), pp. 158–60; Norman Vincent Peale, *The True Joy of Positive Living* (New York: Morrow, 1984), pp. 136–38. That this man, one of two on the Consistory who tried to maintain that body's traditional role of responsibility for the clergy, was a source of great uneasiness to the minister Peale makes very clear in his autobiography. There were two related issues involved: one was Peale's unwillingness to be controlled by a person appearing as an "employer"; and the other was

his longstanding difficulty in dealing with confrontation. Gordon's book, an authorized biography, refers to the event in the context of Peale's problems with "authoritarianism."

10. Syracuse *Post Standard*, January 14, 1933.

11. Milwaukee *Sentinel*, July 5, 1933. Peale's comments also appeared in other newspapers around the country.

On occasion, Peale was given to attacking right-wing leaders as well. He opposed Father Charles Coughlin and supported Roosevelt's proposed veto of the Patman Bonus Bill as a device for reining in "this mad priest." *New York Times*, May 13, 1935.

12. News release, compiled by Wesley A. Stanger, 50 Church Street, October 2, 1934; NVP Ms. Coll. Box 192.

13. In Robert Moats Miller, *American Protestantism and Social Issues* (Chapel Hill: University of North Carolina Press, 1971), pp. 115, 122.

14. Ibid. Miller's work is an excellent study of these issues.

15. Collegiate Church *Yearbook*, 1933, pp. 678–79.

16. "Comparison Every Member Canvass Budget, May 1, 1932 and May 1, 1933"; NVP Ms. Coll, Box 181.

17. Peale, *True Joy*, p. 133; Gordon, *Norman Vincent Peale*, pp. 165–66. In his autobiography, Peale records that he had the very crucial "conversion" experience in Keswick, England, during this summer of 1933. Gordon locates it in 1934. It is quite possible that since Peale produced some of his material from memory rather than notes his dates are slightly off. Since other evidence, essentially letters, indicates that all four Peales traveled together in the British Isles in 1933, and the Keswick experience involved only Ruth and Norman, it seems likely that Gordon's 1934 date is correct. A commemorative pamphlet, produced by Marble Church in 1972 to celebrate the fortieth anniversary of Peale's ministry there, also gives 1934 as the date.

18. On occasion, Peale would use anecdotes about his children in sermons. The documentary evidence that referred to the children was fullest for the summer months and when the family spent weekends and holidays at the farm in Pawling, New York. When family members wrote to each other, the letters they exchanged were brief and concerned mainly with practical matters. In so public a family, interviews and oral history generally provided little additional information.

Ruth was the emotional center of Norman's life, and apparently he of hers. They were devoted parents, but perhaps even more devoted spouses. Summers, for example, Ruth and Norman traveled abroad while the children went to camp. The children regularly attended summer camp, and the two older children, boarding school as well. As adults, they became a very close-knit group.

19. Collegiate Church *Yearbook*, 1934, pp. 88–85.

20. Peale, *True Joy*, pp. 134–35. Because Peale has retold this story many times, it appears in a variety of places—sermons, books, pamphlets, interviews. As noted above, both Gordon's biography and Marble Church's commemorative pamphlet honoring Peale for his forty years as its minister dated the Keswick experience in 1934.

21. "How to Overcome Your Inferiority Complex," unedited first draft, composed with the assistance of Blake Clark, typescript, undated; NVP Ms. Coll., Box 127. This subsequently appeared in February 1955 in *The Reader's Digest*.

22. Allan R. Broadhurst, *He Speaks the Word of God* (Englewood Cliffs; NJ: Pren-

tice-Hall, 1963), examined this technique Peale used. Common within New Thought, Peale used it for all his public presentations and later taught it to clergy at the school for ministers he conducted in Pawling. Oncologists use imaging by asking patients to "imagine" healthy cells, for example, destroying sick or cancerous ones. Athletes who "visualize" themselves in a race claim to be as exhausted at the end of the mental practice as if they had actually competed in the race.

23. Conversation with NVP, Pawling, New York, February 23, 1981.

24. Charles S. Braden, *Spirits in Rebellion* (Dallas, TX: Southern Methodist University Press, 1963), p. 391. In a conversation, Peale referred to his father's description as "Methodist evangelism, Dutch Reformed Calvinism, Unity, and positive thinking." Interview with NVP, Pawling, New York, February 23, 1981.

25. Miller, *American Protestantism and Social Issues*, p. 124.

26. There are many categories of "request mail" in the Peale files. As the number of his conduits to the public increased, people usually wrote in terms of where they had encountered him—the radio program, his books, sermons, *Guideposts*, the newspaper column, and for a few, just his celebrity in general. An occasional postcard, addressed simply "Norman Vincent Peale, New York City," would reach him. Four boxes in the Peale Collection (593, 594, 595, 596) are devoted exclusively to request letters received through the "Art of Living" program in 1937–1938. The writers came from all over the country, although in addition to the metropolitan New York area, the cross-country reach of the Bible Belt predominated. People wrote of their emotional and physical problems and the benefits they derived from Peale's talks. There is little difference in content between these letters and the "problem letters" he received through Marble Church or the Foundation for Christian Living.

27. J. Stillson Judah, *The History and Philosophy of the Metaphysical Movements in America* (Philadelphia, PA: Westminster Press, 1967), pp. 178–82, 183. Also Braden, *Spirits in Rebellion*, pp. 26–45.

28. Norman Vincent Peale, *You Can Win* (New York: Abingdon-Cokesbury Press, 1938), pp. 16, 17. Also Peale, *The Art of Living* (New York: Abingdon-Cokesbury Press, 1937).

29. Peale, *You Can Win*, p. 23.

30. Peale, *The Art of Living*, pp. 118, 119.

31. Peale, *You Can Win*, p. 49.

32. Quoted in Brooks Holifield, *A History of Pastoral Care in America*, p. 221.

33. Quoted in Robert Moats Miller, *Harry Emerson Fosdick* (New York: Oxford University Press, 1985), p. 251.

34. See Robert Powell, "Healing and Wholeness: Helen Flanders Dunbar and the Extra-Medical Origins of Psychosomatic Medicine," doctoral dissertation, Duke University, 1974.

35. *Who's Who in America, 1956–57.* Blanton's obituary in *The New York Times*, October 31, 1966, notes that he received his diploma from the Royal College of Physicians and Surgeons in 1932.

36. Interview with Stephen Pritchard, Director of Training, Institutes of Religion and Health, New York City, June 17, 1981.

37. American Foundation of Religion and Psychiatry, Report, typescript, 1952; NVP Ms. Coll.

38. Stephen A. Hoeller, *The Gnostic Jung* (Wheaton, IL: Theosophical Publishing House, 1982), Prologue, pp. 1–10.

39. Russell Jacoby, *Social Amnesia: A Critique of Conformist Psychology from Adler to Laing* (Boston, MA: Beacon Press, 1975), p. 50.

40. Commissioner of Department of Mental Hygiene (Dr. Newton Bigelow), New York State, to NVP, Oct. 14, 1952; NVP Ms. Coll.

41. Norman Vincent Peale, "Report of the President" of the American Foundation of Religion and Psychiatry, Sept. 20, 1956; NVP Ms. Coll.

42. NVP to Smiley Blanton, Nov. 11, 1949; NVP Ms. Coll., Box 228.

43. Peale, *True Joy*.

44. Interview with NVP, July 5, 1989, Pawling, New York.

45. Ibid.

46. Interview with NVP, July 7, 1989, Pawling, New York. Peale claimed that by 1960 he was spending a third of his time with the business community, and that since his retirement, he spent "all [my] time there," that is, the secular world. His secretary made detailed plans for his out-of-town speaking engagements, and based on a reading of those, it appears that as early as 1945 there were months when he spent from Monday to Thursday on the road at least twice during the month. Notes on Speaking Engagements; NVP Ms. Coll.

47. NVP to Theoda A. Gage, Mar. 26, 1943; NVP Ms. Coll.

48. Interview, NVP, Pawling, New York, July 5, 1989.

49. Interview with NVP, July 7, 1989, Pawling, New York.

50. Peale, *True Joy*, pp. 226, 227.

51. Untitled Resolution of Collegiate Consistory, June 4, 1943; NVP Ms. Coll., Box 201.

52. There is a fairly large file about this event, an indication of its seriousness to Peale, and perhaps to the church. It encouraged him to consider resigning from Marble. Special Work Committee of Marble Collegiate Church (M. F. Stunkard) to Dear Friend (of Marble Church), Oct. 3, 1941; D. J. Stockbridge to NVP, Oct. 5, 1941; "An Analysis of the Contents of the Communication by Dr. Macy to the Classis of New York, n.d.; NVP to Dear Dad, Jan. 9, 1942; Clerk of Collegiate Church (Henry P. Miller) to NVP, June 4, 1943. NVP Ms. Coll., Boxes 185 and 188.

53. Smiley Blanton to NVP, July 18, 1940; NVP Ms. Coll., Box 180.

II

The Movement

There was nothing radically new in the distinctive element in Peale's message, the emphasis on positive thinking. As an aspect of the religious science-mind cure tradition, it had been a cultural potential in American life since the mid-nineteenth century. One factor that enabled positive thinking to move from its position on the social and intellectual margins to the mainstream was Peale's ability to harness the resources of modern life to the transformative process within his own ministry.

Peale's organization—the movement—he liked to describe as a structure of ministry that was supported by five "pillars." These pillars were the Marble Collegiate Church, *Guideposts* magazine, the Institutes for Religion and Health, his many books but especially *The Power of Positive Thinking*, and the Foundation for Christian Living. If these supports were found in the secular world they might be described as the interdependent agencies of a substantial industry. In Peale's case, with the exception of Marble Church, and to some extent the Institutes for Religion and Health, these agencies formed the basis for his independent public ministry. The nucleus of it started in the 1940s, it assumed a mature separate existence after 1952 as a result of the public success of *The Power of Positive Thinking*.

It is impossible to imagine what might have become of Peale's movement of practical Christianity without his celebrated book. Its message, its sales, its controversial celebrity, formed the centerpiece of the movement. The book made Peale a public personality nationally and internationally, ensuring his access to new opportunities that meant growth for the other "pillars" of his ministry. The publicity accorded his book had ramifications for *Guideposts*, which experienced a burst in sales along with a change in format and editorial direction. His visibility increased his ties to the business community, where he gained new contacts and additional support for other ventures.

The headquarters for his personal ministry was the Foundation for Christian Living, renamed the Peale Center for Christian Living. Essentially a publishing and communications hub for his work, the Center managed Peale protocol and consequently played the important role of helping to set his priorities. That Ruth Peale became the chief executive officer of the Center was significant: A hands-on executive rather than a symbolic figurehead, she was an influential member of what Peale accurately described as a team ministry.

CHAPTER 4

"More Than a Magazine": The Guideposts *Story*

> We cannot overstate our debt to the Past, but the moment has the supreme claim. The Past is for us; but the sole terms on which it can become ours are its subordination to the Present. . . . The divine gift is ever the instant life, which receives and uses and creates, and can well bury the old in the omnipotency with which Nature decomposes all her harvest for recomposition.
>
> Emerson
> Essay on Quotation and Originality
> 1859

From 1942 on, Peale blazed about the country at a pace so frenetic his Pawling, New York, neighbor Lowell Thomas called him the "jet propulsion man." Through contacts in religion, politics, business, and industry, he was creating circles of supporters that collectively formed a vast national network, essentially the constituent part of the Phenomenon of Pealeism. It was indeed a phenomenal creation: Like an octopus, it gradually developed tentacles that reached deeper into new areas of popular culture. The heart of this vital organism after 1950 was the Foundation for Christian Living (FCL), the center of a thriving religious empire.[1] The FCL was the product of Ruth Peale's creative direction of the sermon publications project at Marble Church. It quickly bore her personal stamp, evident in its efficient, innovative, and culturally sensitive style. The link between the center and the large national constituency was a vital and complex communications network formed by Peale books, published sermons, and regular FCL mailings.

The most influential disseminator of the Peale message and its symbol system, however, lay elsewhere, in *Guideposts* magazine, the compact, inspirational monthly publication modeled after *The Reader's Digest.* It was the first and for a long time the only link tying the various parts of Peale's network together. More than any other aspect of his ministry, *Guideposts* gave

Pealeism an easily understood public identity. Politically, early *Guideposts* found a home for the potentially antinomian message of positive thinking within Cold War conservatism. Culturally, it identified the beleaguered middle-class victim of the crush of modernism as its archetypal reader. Therapeutically, it offered examples of and remedies for coping with the crises of modern life. A newcomer to the tradition of positive thinking-inspirational literature, *Guideposts* recognized a need that no other agency, publication, or mainstream organization was addressing.

At its founding, Peale had a vision of what the magazine should be: "an organ for a great, positive Christian movement," he announced. The movement would be predicated on a call for a return to simpler times. It was a metaphorical bid for a return to an early time when presumably people were more conscious of their religious responsibilities, and when, politically, the nation seemingly honored a special and sacred calling. The "simple" message of Christianity, Peale said often, had been distorted by an official religious bureaucracy, with his own mission part of an effort to restore a fractured heritage for church and state. Politics were as important as religion in his plans.

Consequently, the editorial philosophy of *Guideposts* at the time of its 1944 inception reflected Peale's civic and religious priorities. Celebrated in his various support networks, they were best summarized as the ideas of Americanism, free enterprise, and practical Christianity. Integral parts of his message, Peale promoted them actively in the early 1940s in his parochial ministry and in extended contacts with several partisan political groups. From 1935 to 1952 three conservative anti-New Deal/Cold War political organizations claimed his time and his enthusiasm: the Committee for Constitutional Government; Spiritual Mobilization; and the Christian Freedom Foundation. For a while, too, he cooperated with H. L. Hunt's Facts Forum. His exposure to these political lobbying groups—which described their efforts as "educational"—convinced him of the persuasive power of popular literature. *Guideposts* magazine emerged out of this politically charged environment.

Like the counseling clinic Peale started at Marble Church, *Guideposts* went through several transformations, with growth the important constant, sometimes breathtakingly so. Intended to have the religious and business communities better understand each other, it began in the 1940s with the expansive accounts of public men, whose sources of inspiration resembled Peale's own. The Bible, the McGuffey Reader, and Horatio Alger stories were the background references. In the beginning, the paper depended on the support of industrialists, and they in turn distributed free copies to their employees. But by 1952, when *The Power of Positive Thinking* topped the best-seller list, a new *Guideposts* audience began to define itself. No longer did male industrialists and their employees predominate; the new clientele was increasingly female, middle-class, and middle-aged. The adaptive quality of Peale's message and the fluid structure of the paper made it possible for the publication gradually to respond to a self-defined readership that changed as it grew over the years.

The paper tried to give life to its motto, "More than a magazine," by cultivating the image of the "*Guideposts* family." Its most successful and enduring service in that regard was its national prayer fellowship. It developed a number of other outreach services, but the prayer network was the centerpiece of the operation. Organized on the model of Silent Unity, it was intended as a personal bond between the journal and its audience. It was the prayer fellowship that ideally would transform the paper from just another inspirational periodical into the "organ for a great, positive Christian movement," changing readers into intimate members of the "*Guideposts* family." Besides using the printed prayer calendar, readers were encouraged to send requests for personal prayer to the paper, where their concerns were guaranteed to be addressed in the weekly staff prayer meeting. The paper invited subscribers to correspond, and once an exchange was established, it became an ongoing conversation between reader and staff, an interaction that paradoxically became even closer when the introduction of new computer technology allowed for personalized responses at the touch of a few keys.

Although the readership since the early 1950s has been predominantly female, the prototypical target reader over the life of the paper has remained a metaphoric male. In the early years of publication, the model of maleness was a vigorous Cold Warrior, in part a reflection of Peale's desire to portray practical Christianity as virile and robust. Gradually that image changed, in keeping with changes in editorial philosophy. By 1985, when *Guideposts* editor Van Varner characterized the symbolic male reader, he described him as a "lonely salesman in the motel room, away from his family and his spiritual home."[2] In a curious way, the description could have applied to Peale himself—a tireless traveler, constantly on the road, a solitary journeyer until in advanced years his wife was able to accompany him. The quality of homelessness on the most obvious level also reflected the paper's consistent story line. In manifold ways it told the saga of the single individual, brought face-to-face with a crisis of modern life and eventually triumphing over it. Peale once said that the secret to the paper's success could be found in six words, "Find a need and fill it," although in the early years of publication he and the staff had difficulty identifying a need with a particular audience.

In 1945 when the paper was launched, Peale believed the need of the hour was for a Militant Church, its members galvanized to fight Communism and preserve free enterprise. Over the next decade, *Guideposts* readers began to identify their own need, and it was a more personal one, connected to the shift in the demographic profile of the subscription list. The emergence of a large and growing audience of women persuaded the staff to temper Peale's bullish politics with more existential issues of generic despair, offering literary explorations of that metaphoric quality of homelessness.

Homelessness was a telling social term, then and now. It reflected not only physical alienation but also the psychic and spiritual rootlessness of modern life. The redefined *Guideposts* reader was not presumed to be one of the destitute horde who crowd urban doorways—though the conditions were symbolically related. Rather, he or she seemed to have more in common with the sufferers of the kind of psychic homelessness that sociologist Peter Berger regarded as part of the now familiar fallout of the march of modernization: the disruption of social, familial, and geographic roots and the loss of regard for an historical past.[3] The readers were people who felt dispossessed from their traditional surroundings, their symbolic homesteads, and forced to cope by their wits, they found themselves frequently foundering rather than coping. Every *Guideposts* story was a tale of personal crisis, its implicit message that with the right skills and relevant knowledge each person could surmount his or her particular trial and lay the foundation for a new psychic homestead.

"There Is Absolutely Nothing Political about Guideposts*"*

Guideposts was not conceived with that specific therapeutic intent. It took root in the overheated soil of Cold War politics, its seeds planted at a time when Peale's personal ministry was expanding, and his political interests were richly nurtured by the partisan political groups that claimed his attention. At the time he was developing ideas for the paper, Peale was much preoccupied with partisan issues, serving as chair from 1942 to 1945 of the Committee for Constitutional Government, a conservative anti-New Deal citizens' lobbying effort, while simultaneously contributing to the grass-roots, free enterprise program of California clergyman James Fifield, called Spiritual Mobilization. *Guideposts* was envisioned as a penetrable barrier between the two communities, the religious and the politico-economic, where Peale had ministries. From Fifield he learned the value of a grass-roots communication network, and from his service with the Committee for Constitutional Government, particularly its effective mass mailing techniques, he acquired important information about technology and public relations. Edward Rumely, Executive Secretary for the Committee for Constitutional Government, whose political past made him suspect to liberals, was an astute strategist in planning Committee campaigns, and he made his abilities—and some of his mailing list—available to Peale as he prepared to launch *Guideposts.*

When advance notices of the *Guideposts* project finally went out to the papers, appearing in publications as diverse as *The New York Post* and the Methodist *Zion's Herald*, they drew immediate responses, from the curious as well as the critical. Critics were especially leery of an organ associated with the right-wing views of Spiritual Mobilization and the partisan Committee for Constitutional Government. Opponents pointed not to Peale's theology and its possible grass-roots' dissemination but to his politics, and their comments were decidedly unfriendly. Dr. Dwight Bradley, a clergy member of the leftish National Political Action Committee and a continuing nettle in Peale's public life, believed that Peale's claim that the paper would be "decidedly religious" was untrue, his explanation of its purpose devious, so that "every effort should be made to unmask this campaign [to develop the paper], not because it is reactionary, but because it pretends to be purely religious."[4] Bradley's concerns, he admitted, were related to the "great political conservatism" of the paper's sponsoring committee, a charge related to "Dr. Peale himself, to Dr. Fifield of Los Angeles, to [newspaper publisher and Committee founder] Mr. Frank E. Gannett and to Captain Eddie Rickenbacker."[5] If Peale's profession of dedicating the paper to a religious revival was true, Bradley remarked, it would finally "free his ministry . . . from the compulsions of redbaiting."[6] Peale reminded Bradley that the magazine "*is* purely religious," as he could verify when the first copy appeared.

But liberal watchdogs were not assuaged. Among the sharpest critics were two vigilant editors—George Seldes, pioneering critic of the commercial press and founding editor of the weekly prolabor *In Fact*, and Guy Emery Shipler, editor of the liberal Episcopal periodical *The Churchman*. Seldes, suspecting a conspiratorial link between *Guideposts* and the Committee for Constitutional Government, said in a letter to Peale: "As for *Guideposts*, I assure you that if your name had been attached to this organization alone, and not to the political, lobbying and sinister—and these words are from the Congressional Record—Gannett outfit, I would never have taken up this matter."[7] Shipler framed his challenge directly: "When you bring suit against George Seldes for what he presented in such issues of *In Fact* as those of May 26 and April 2, 1945, and win it, I shall be convinced *The Churchman* is misled."[8]

Some of his far less acerbic clerical colleagues also had reservations about the paper. They had been alerted to Peale's partisan sentiments through regular press coverage of the political content of his sermons and by reports of his fellow-traveling with cronies whose conservatism bordered on reaction. Two out of the four local clergymen he approached for endorsements had misgivings. To his old friend John Sutherland Bonnell, Peale offered the reassurance that "There is absolutely nothing political about *Guideposts*." Apparently persuaded, Bonnell contributed an article to the first issue.[9]

His New York neighbor Ralph Sockman, minister of Christ Church, Methodist on Park Avenue, had more serious reservations, which were somehow circulated to Seldes' *PM* magazine. There it was reported that "Sockman talks of quitting group that was formed to fight collectivism." To allay Sockman's fears, Peale told him that "you have my definite word that *Guideposts* is not and will not be in any sense political." He had resigned his post as chair of the Committee for Constitutional Government, he told Sockman. He hoped that his colleague would remain on the advisory board of *Guideposts*, yet he conceded the possibility that Sockman could "feel embarrassed" by continuing the association.[10]

Norman's old friend, Methodist Bishop Herbert Welch, wanted to support the project, but he, too, had doubts. After looking over some preliminary documents describing the new paper, the Bishop told Peale that he could understand how some observers gained the impression that "this was a movement for economic and political conservatism, or even reaction."[11] Advance circular letters describing the magazine were sent on Peale's behalf by conservative Republican Congressman Ralph Gwinn. A spokesman for a Jewish agency, fearing anti-Semitism, asked if *Guideposts*' supporters would "resign from a country club because its membership is 'restricted.'"[12] Victor Riesel, a regular monitor of Peale activities from his vantage point with *The New York Post*, joined his voice to the chorus of critics who worried aloud about the political credentials of *Guideposts*' wealthy friends.

Not surprisingly, *Guideposts*' angels were conservative men of substance and entrepreneurial success: Eddie Rickenbacker, war hero and president of Eastern Airlines; Branch Rickey, President of the Brooklyn Dodgers baseball club; Walter Teagle, Marble Church member and former chairman of the board of Standard Oil Company of New Jersey; Stanley Kresge, the president of S. S. Kresge; W. N. Grant of Grantville Mills in Grantville, Georgia; Howard Pew of the Sun Oil Company; and Peale friends Frank Gannett, E. F. Hutton, and Lowell Thomas. Pew, a right-wing evangelical Presbyterian, would be a continuing source of support for Peale. His political associate from Spiritual Mobilization, James Fifield, participated by raising money and subscriptions on the West Coast. Many of these early backers had assisted the magazine precisely because they believed they could count on Peale's political commitments to counteract any left-wing bias in the two target audiences, the church and industry. Peale therefore had to tread a fine line, promoting the project to them while defending it from liberal critics. Ever eager to gain acceptance, he struggled to find the right words for each group.

He did not always succeed, although he tended to find it easier to explain his work to friends in industry than to fellow members of the clergy. On occasion Peale referred to himself as a "willy-nilly man," seeming to sense that his desire for approval sometimes led him down dark alleys. Mistaking his announced intention—of creating an organ for a great Christian movement—many of his industrial supporters took for granted that *Guideposts* was simply an extension of Peale's political goals, with his reassurances to them convincing them they were right. To his conservative friends, Peale spoke of *Guideposts*' ability to "refertilize the soil of American life by widely spreading religious ideas, [so] we can counteract the communistic virus . . . and defeat left-wing influences such as Communism."[13] In a different context, he argued that *Guideposts* might be compared to an inoculation, which, circulated among exposed groups like teachers and workers, could create a "natural resistance against communism or socialism."[14] He liked Edward Rumely's phrase that "freedom was the child of religion," and he tried to work it into his appeals whenever he could.

He tapped into the fears of businessmen when he noted that many "clergymen have gone over to the anti-business and collectivistic point of view" and that a *Guideposts* contribution might be usefully directed to circularizing them with the paper.[15] In answer to the concerns of a Georgia textile manufacturer, Peale pointed out that Ralph Sockman was not "left-wing," was not a "parlor pink," nor was he, in any case, any longer a part of the *Guideposts* advisory board.[16] He reported to Rochester publisher Frank Gannett, one of the four largest original contributors to the paper, that since the clergy were being propagandized by "left-wing CIO-PAC forces," there was a special need to "get our project [*Guideposts*] under way [so] we can exert a constant and widening influence in the right direction."[17]

While one of Peale's motivations for the magazine came from his awareness of the popularity of religious science literature, another derived from his work with the Committee for Constitutional Government. He had learned from his Committee experience how effective a direct mail lobbying effort could be in shaping public opinion. He never conceived of *Guideposts* as existing in a political vacuum, but rather serving as the communications link that tied his whole ministry together. Peale confided to a friend that he hoped *Guideposts* might be the means for retaining the loyalty of Protestants being enticed away by sects such as "Christian Science, the Oxford Group, Unity, and similar organizations."[18] Feeling exasperated by the persistent criticism made of the paper, he asked "Why, in the name of heaven, has everything today become so 'sinister'?"[19]

But Peale's continued identification in the popular press as part of the anti-Communist vanguard encouraged his critics to regard *Guideposts* as yet another reactionary Cold War broadside, all the more dangerous for its seeming disguise as a legitimate mainstream point of view. That Edward Rumely, a man convicted, and acquitted, of trading with Nazi Germany, remained a behind-the-scenes strategist for the paper's planning only fanned the flames of controversy. Still executive secretary of the Committee for Constitutional Government at the time, Rumely participated in planning for *Guideposts*, not only sharing his extensive experience in direct mail and market strategies but offering his home for editorial discussion sessions and sharing subscription lists.[20] Rumely's practical know-how enabled him to anticipate the rate of positive and negative returns, the cost of paper, and the need for crucial timing. Equally important, he had mailing lists, names and addresses of contacts from which the planners drew in their initial appeals.[21] In-house correspondence for the paper was sometimes on the letterhead of the Committee for Constitutional Government, and *Guideposts*' financial sponsors, some of whom were also members of the Committee, occasionally received announcements about both organizations in the same envelope.

Rumely gave generously of his unsolicited counsel. He suggested, for example, that injecting a "materialist argument" in an otherwise religious letter might prove productive in circularizing southern clergy in particular with *Guideposts* subscriptions, and that it would be far cheaper to make recordings of some articles and play them on the radio rather than rely exclusively on mail distribution.[22] And because Rumely had high expectations for an initial subscription run of 100,000 a month, he advised Peale to make *Guideposts* his top priority, to "eliminate all other engagements" temporarily as he took on "the most difficult editorial task imaginable."[23]

The businessmen close to the planning quickly recognized the practical potential inherent in a magazine promising "sound economic implications," particularly if it could be given to "employees in those localities where the radical thinking is worst."[24] Peale himself drafted an invitation to clergy to support the project, sounding a restorationist alarm: "Cannot the spiritual power of Christianity be mobilized and brought to bear upon the problems that confront us?" just as the Communists mobilize political power?[25]

As with many other Peale developments, *Guideposts* emerged from the two dominant influences in his thinking, his practical Christianity and his political conservatism. His politics often parted company with the Emersonian political tradition, which was alternately reformist and apolitical. With Peale, his early exposure to populist strategies had had a formative effect. It could as easily have been said of him, as it was allegedly said of President Andrew Johnson, that he longed to be a snake in tall grass that he might snap at the heels of passing wealthy planters. His version of populism, which wedded practical religion and conservative politics, consistently found its mark in the programs of the nation's liberal leadership. For almost its first decade of life, *Guideposts* reflected the original influences that gave it life: the Cold War environment and the political angels and business associates who sponsored it.

At the time, *Guideposts* also failed to realize Peale's vision of advancing "a great, positive Christian movement," because its assertive Cold War jeremiads and expansive entrepreneurial testimonials did not address the needs of his popular audience. To them, Peale was identified not with the Manichean politics of the postwar era but with a softer, therapeutic Christianity spoken in the language of evangelicalism. Swimming with the sharks in the reactionary waters of postwar McCarthyism, he was nearly eaten alive—political watchdogs reported his every move, on occasion to his great embarrassment. When Peale abandoned that contest and returned to a less activist Republican conservatism, his harmonial audience for *Guideposts* increased, as readers looked for positive techniques for gaining self-empowerment.

"Don't Think Lack"

A shoestring operation in its early years, *Guideposts'* future success was then so well camouflaged that even Peale had trouble envisioning its extraordinarily bright future. First published in 1945, its neophyte staff was a makeshift team consisting of Peale, some secretaries, and publisher Raymond Thornburg. It was missing an editor until the addition of Grace Perkins Oursler, a Roman Catholic and the wife of *Reader's Digest* editor Fulton Oursler. Her hiring emphasized the affinities between the *Digest* and *Guideposts* and Peale's hope to create an ecumenical audience. Editor-in-chief Peale had the tasks of soliciting articles from friends, establishing an editorial philosophy, and raising money to keep the paper operational. When funds ran low, he took his tin cup and went on the road. At one particularly gloomy staff meeting, according to a classic story in the Peale repertoire, as the group contemplated a mountain of bills and very limited receipts, the session brightened with new life when visiting benefactor Tessie Durlach advised the assembly "not to think lack"—not to dwell on what they did not have but to envision a future holding bulging files of subscribers' names.

Publisher Raymond Thornburg was Peale's Pawling, New York, neighbor, founder of the Pawling Rubber Company and an old friend from college days. Thornburg's wife was Pherbia Thomas, sister of newscaster Lowell Thomas, another resident of Pawling, who in 1944 convinced the Peales to buy a small farm for a country retreat on prestigious Quaker Hill in the Dutchess County village. Lowell Thomas would remain a close friend and Peale supporter over the years. Consistent with Peale's expectation that the magazine would be a journal for "ordinary" people, Thornburg thought it appropriate for it have a small-town address and consequently arranged for an office over a store in Pawling.

The original design for the paper was an eight-page fold-out pamphlet, modeled on the *Kiplinger Letter.* Like *The Reader's Digest,* it was a human interest periodical, only more inspirational, its aim to show how practical Christianity "worked" in daily life. To that end, Peale solicited stories, always first-person accounts, from men in business, sports, and the professions, in which the writer told of the "helpfulness he has received from his religious faith."[26] In the first five years of the paper's existence, the articles tended to be preachy, heavy-handed accounts by men with public reputations telling about the life-changing results of a personal spiritual encounter with God. Rife with fears of communism, the pieces also extolled the laissez-faire individualism of the authors in orotund, Horatio Alger style. Peale envisioned an ecumenical readership of Protestants, Catholics, and Jews, although it was inevitable that Peale's expanding public presence would create an audience almost exclusively Protestant. In time there was a small list of Catholic subscribers, but a Jewish constituency never really materialized.

Sold in the early years through limited individual subscriptions and primarily through group orders to churches, schools, and business organizations, *Guideposts* acquired a subscription list of 55,000 and a twenty-four–page format by 1948, just a year after a devastating fire had destroyed its files and mailing lists. After the fire, Lowell Thomas had broadcast an appeal to subscribers to write in with the necessary renewal information, and as a result *Guideposts* not only recovered its original readership but added substantially to it. Hoping for a heavy circulation among the clergy, business leaders, salesmen, and industrial workers, Peale deputized old friends and family members to solicit sales, ideally corporate ones, from these groups. He would go himself to cities around the country where *Guideposts* luncheons had been arranged for him, taking advantage of promotional opportunities planned by local businessmen.

The practical result of this strategy of building sales through bulk subscriptions to employers was that most readers were, in the words of Editor Oursler, "for-free" readers, whose loyalty could be quite limited. She feared that many of them, receiving the magazine from their companies, would regard it "either as propaganda or as goodie pap," and she hoped that Peale would find it possible to make editorial decisions independent of the feelings of his industrial supporters.[27] "I do not hope for great enthusiasm from people who have a magazine bought for them instead of choosing it themselves," she said, and as potential independent subscribers they would be "a tougher audience to win here than anywhere else."[28] Eager to avoid offending her sensitive boss, yet mindful of his susceptibility to appeals from his conservative friends, Oursler observed to Peale that it was rumored he was "a prize worrier and something of a shilly shallier and an easily swayed person," a comment she then qualified by explaining that it was a surface appearance, not found in the "inner you."[29] She suggested that he and the staff begin a re-examination of editorial philosophy, and a campaign that might lead to a 2 to 5 percent individual renewal rate once employers stopped free distribution.

In these early years industrial subscribers kept the magazine going. They not only bought copies for their own employees but in some cases purchased them for teachers in neighboring school districts, as well as for local churches. The star performer within industry was the R. J. Reynolds Tobacco Company in Winston-Salem, North Carolina, where 12,000 copies monthly made that southern city the area of greatest penetration for the magazine at the time. In-house calculations at *Guideposts* estimated that in every other home in Winston Salem a person received the paper. Some companies with strong unions, sensing that employees would equate free distribution with paternalism and manipulation, turned down overtures from *Guideposts.* General Motors Corporation managed a compromise with employees, ordering enough copies for its reading racks—sometimes as many as 125,000—for each person to have one. A friend of Peale's ordered 1,500 for supervisory personnel at Edison Industries in New Jersey; Eddie Rickenbacker distributed 8,000 among Eastern Airlines workers; and Conrad Hilton had copies placed in guest rooms at selected hotels in his chain. Peale estimated that in 1947, 26,000 industrial foremen were receiving copies of the publication.[30]

It was primarily Peale's personal popularity that boosted sales. His national audience showed a dramatic increase as, overcoming an early fear of flying, he found it possible to take his message to three or four widely scattered audiences—church groups, sales and business conventions, civic organizations—during the course of a single week, reaching areas previously neglected. Thanks to the efforts of Ruth Peale and Sermon Publications, Peale's sermons went to 100,000 recipients by 1950, while his "Art of Living" radio broadcast generated 6,000 to 8,000 letters a week, and his syndicated newspaper columns made yet another contact with his public. *The Reader's Digest*'s publishers, DeWitt Wallace and Lila Acheson Wallace, already Peale fans, became equally enthusiastic about *Guideposts*—which was located not far from the *Digest* headquarters in Carmel, New York—and allowed the infant publication to read manuscripts it had rejected. Hearst Publications ran excerpts from the magazine in some of its Sunday syndicated offerings. But it was Peale's speeches to business groups that kept the bread on the table for *Guideposts*, and he was spending more time on the road to stay in contact with potential contributors.

His extraordinary rapport with these business leaders enabled the paper to chalk up a lengthening list of industrial subscribers. A Peale speech to a packed audience of delegates to the annual meeting of the National Association of Manufacturers in New York in 1949 prompted one enthusiastic listener to begin, spontaneously, to generate a list of potential subscribers to be forwarded to the magazine. An almost similar outcome followed his talk the previous month to the annual meeting of the National Coal Association.[31] Peale wrote to Congressman Brooks Hays in 1949 that he had become "only a figurehead" at the paper, his job to raise money and ensure its financial future through personal contacts.

Although Peale retained closer supervision of *Guideposts* than he did of the Institutes of Religion and Health, his executive management style in both instances was generally similar, allowing the professionals on the staff to make the daily decisions while retaining the executive veto for himself. At *Guideposts* that led to periodic re-examination of editorial philosophy: How to continue to tell the Peale version of practical religion, utilizing first-person testimonials about the spiritual strength gained through overcoming crises, while still producing a modern, sophisticated, well-written monthly that appealed to independent readers.

Grace Oursler had focused the problem when she noted the difficulty of converting "for-free" readers into paid subscribers. In 1948 the paper was receiving only twenty-five reader letters weekly, with one-third of that small number critical. To win new readers, Oursler recommended editorial change. Peale's column, appearing in every issue, usually a ghostwritten composite of his sermons, sold the magazine, she believed, and had to stay. But in her editorial opinion, the clergy generally was too much present in each issue, and its presence should be reduced. Instead, she suggested, there should be more articles by and about women. She also endorsed the "how-to" format found in almost all articles: "The average reader," she observed, "wants definite dos and donts." And for the time being, its conservative political roots would remain undisturbed. The greatest number of reader requests for a reprinted article in 1948 was for a piece by ardent Cold Warrior J. Edgar Hoover.[32] In an editorial response to a reader who canceled his subscription because he was "shocked at your Communist attacks," the paper commented: "We print and shall continue to print Communist exposures because Communists are Anti-God, Anti-Christ. . . . *Guideposts*' policy is in no way political."[33]

In 1951 an ailing Grace Oursler retired from the paper, her place taken by Leonard LeSourd, whose long tenure, until 1974, had a formative influence on the publication. When Oursler left, the periodical was enjoying modest, though uneven, success, due largely to the nature of its subscription list, which was heavily weighted in favor of industrial orders.[34] A change in subscriptions from a major supporter—for example, U.S. Steel, which ordered 100,000 copies in 1952 and then cut it to 30,000 two years later—made the paper extremely vulnerable to market or ideological shifts. LeSourd tried, with some success, to reduce that vulnerability during his long term as editor.

Peale had initially identified a market for *Guideposts* that was politically tough and commercially soft and that was disconnected from his larger audience of religious seekers for answers to personal crises. The very limited success of *Guideposts* in the forties was also a commentary on the Cold War, the self-made men who advanced it, and the vast but silent segment of the evangelical-metaphysical population that related to it in only a marginal way.

A New Image for Guideposts

Leonard LeSourd was a staff writer, a young ex-pilot, when he took over as editor of *Guideposts*, a position he was to hold for twenty-four crucial years. He had high journalistic standards and encouraged authentic reporting—even disguising himself as a bum on one occasion and living on the streets to report on conditions among the poor—yet he was also described by colleagues as a "spiritual" man who maintained the tradition of Monday morning staff meetings devoted to answering prayer requests. During his term as editor, he married Catherine Marshall, widow of the popular preacher Peter Marshall and a prominent author in her own right. His special blend of journalistic skill and quiet spirituality, combined with a toned-down version of anti-Communism, increased the paper's appeal to readers at a time when Peale's popularity soared and the national religious revival created a potentially larger market. He groomed the magazine for its take-off.

If there were any doubts that Peale's popularity sold *Guideposts*, they were banished when *The Power of Positive Thinking* appeared in 1952. On the eve of publication of the book, subscriptions stood at 200,000, built around a soft industrial base. By 1953, two years later, in-house estimates placed the number of subscribers at 500,000, an increase only partly attributable to the

U.S. Steel request for 100,000 copies. At the start of the new decade, in 1961, *Guideposts* boasted a million subscribers, over 40 percent of whom had purchased the magazine for themselves.[35] The decline in industrial sales, reportedly from over 80 percent of total subscribers when the magazine began, to 25 percent some fifteen years later, was the most dramatic indication of change.

Combined with the great popularity of Peale's book and the rising tide of religious revivalism, journalistic changes at the paper contributed to its acquired reputation as a "minor miracle."[36] LeSourd seemed as committed as Peale to seeing *Guideposts* as the hub of a national spiritual network, referring regularly in the paper to the "family" that formed the prayer fellowship. In the first-person stories that remained the distinctive feature of the publication, writers uniformly attributed their ability to deal with crises to the power of affirmative prayer. That was also the central message of *The Power of Positive Thinking* and a constant in Peale sermons and speeches. The paper's anti-Communist stance remained, though more moderate. The self-help, or "how-to" message, now a standard feature, became more descriptive, whether contained in a Peale column about "What to Do about Anxiety" or a woman's story about "How I Survived a Hurricane." Peale's longstanding interest in metaphysical subjects was evident in topical articles on subjects as diverse as how to live harmoniously with the forces in the universe or views on the Science of Mind, spiritual healing, and Moral Rearmament and on J. B. Rhine's work in parapsychology. Traditional middle-class values, about parental roles, sex roles, the work ethic, Sabbath observance, and the like, were woven into the fabric of most stories, their message explicit though not usually overbearing.

Peale's national celebrity actually proved to be something of a mixed blessing for the paper. It sold the magazine, but it kept him from regular attendance at staff planning meetings, despite the fact that he made the judgment on final copy. His absence gave LeSourd greater editorial freedom, the exercise of which brought him into conflict with his boss. LeSourd, in any case, could be only limitedly innovative since there were mandated categories to be covered in each issue: business, self-help, sports, adventure, family problems, and a celebrity feature. From time to time Peale threatened to resign his position with *Guideposts*, usually when tensions converged after a pounding from the press, and then reconsidered. He did not resign, but neither did he become more attentive to the issues at the paper, leading LeSourd to conclude that *Guideposts* was in Peale's priorities "low man on the totem pole."[37]

LeSourd utilized what editorial independence he could extract from the circumstances, the result being a magazine with an obviously different emphasis. Some changes were technical and external. In 1953 Guideposts Associates Incorporated bought the former site of Drew Seminary for Girls in Carmel, New York, for its production headquarters, its editorial staff subsequently moving to New York City offices. After moving to Carmel and experimenting with the latest in mechanized production, packaging, and mailing of bulk orders, the paper became a technically sophisticated operation. Another advantage resulting from its new location in an idyllic community in the hills some sixty miles north of New York City was its even closer proximity to *The Reader's Digest*, with which it shared ideas, manuscripts, and staff. It became a familial relationship when the younger Peale daughter, Elizabeth, married a man then a member of the *Digest's* senior editorial staff, John Allen. The changes in the technical production resulted in alterations in the paper's format, most noticeably when the length increased in 1964 to thirty-two pages. As significant as these modern mechanical innovations were, however, they were ultimately less consequential than the internal substantive changes in the paper's emphasis.

Guideposts continued to run the expansive accounts of entrepreneurial success limned by the heroes of business. But they no longer provided the distinguishing feature of the paper, however, as human interest stories written by those "ordinary" men and women so central to Peale's approach filled an increasing portion of every issue. Initially free-form, the stories rather quickly began to follow a formula: They were all first-person, present-oriented, crisis-centered testimonies to personal triumph over tragedy, achieved through prayer, positive thinking, and common sense. The formula held, whether the writer was a politician, an athlete, or a housewife. The formula was said to "work" in treating the functional debilities described in the accounts in much the same way that orthodox medicine worked, although a more relevant model was the Twelve-Step approach of many behavioral modification programs. In the years LeSourd directed the paper, the human interest accounts became its distinguishing feature.

Two examples taken from issues sixteen years apart indicate the paper's journalistic maturation as well as the increasingly personal and intimate style of the articles. The first appeared in 1952, when LeSourd was just settling into his job, and the other in 1968, by which time LeSourd had placed his signature on the magazine.

In 1943, actress Jane Froman had been injured when the plane carrying her U.S.O. entertainment troupe crashed off the coast of Portugal, killing twenty-four of the thirty-nine people on board in a landing at sea. She told her story in the April 1952 issue of *Guideposts*:

> Conscious, but unable to help myself due to injuries, I was kept afloat by Captain John C. Burn, co-pilot of the plane, who held on to me in spite of his own head and back injuries. After a rescue launch pulled us out of the icy water, they found both my arms broken, one almost completely off; my right arm was fractured and several of my ribs had been cracked. . . .
>
> The doctors didn't tell me how hopeless they considered my case. In a way I sensed their unspoken verdict. "At best she will be a helpless cripple." All except God and myself thought I was finished. . . .

> I began to see how trouble and privation fit into His plan for man's development. To make personal progress in character building, human beings must give of themselves by helping others; therefore my own need became a necessary part of somebody else's development. . . .
>
> In the years which followed, in and out of hospitals, with operations and reoperations interrupting my work, I had little time to feel sorry for myself. Huge doctor and hospital bills accumulated—I had to earn money. . . . The only way I knew was with my voice. . . .
>
> Many months after my accident, I saw for the first time since leaving Lisbon hospital the man who saved me from drowning that night, Captain John Burn, the Pan American pilot. It is hard to describe my happiness when we were married in 1948—or the thrill when, a short time later, I threw away my crutches for good.[38]

Not all accounts ended quite so felicitously, although all stories illuminated the basic themes: heightened self-awareness; the mobilization of spiritual resources, particularly affirmative prayers; and the implementation of self-empowerment.

It was another way of restating the familiar Peale slogan, namely, "picturize, prayerize, actualize." Related to the technique of "imaging" he used in developing sermons, it was based on creating a mental image of a desired goal, praying for the resources to reach it, and finally taking action to bring it about. Success was the inevitable outcome, whether the example was part of a Peale sermon or a *Guideposts* anecdote, with success sometimes measured in terms of an individual's ability to live with imperfection rather than overcoming it. It was uncomplicated spirituality, owing more to the holistic concepts of psychosomatic medicine and behavioral modification than to traditional forms of systematic theology. *Guideposts*, in fact, was not considered a "religious" publication at all, neither by its own standards nor those of its journalistic peers, but rather an "inspirational" or spiritual magazine. It offered an unorthodox approach to religion and healing, one that defined sacred time as that time when the reader received and practiced the rituals described in *Guideposts*, and relocated sacred space in the quiet sitting room where the subscriber studied its words.

The altered demographic profile of the *Guideposts* reader in the fifties came as a surprise to Peale, who had difficulty adjusting to the new reality. The evidence for the change came from a questionnaire the magazine sent in 1957 to 1,000 of its subscribers, and received in return 130 responses. The random survey revealed that women outnumbered men in the readership by two to one. Additionally, it showed that 90 percent were married, most between the ages of fifty-one and sixty, with 50 percent Methodists, followed by Presbyterians, Baptists, and miscellaneous; only one reader was identified as a Roman Catholic, and none was Jewish.[39] The evidence was frustrating to Peale, who had been on the speaking circuit trying to attract industrial supporters, arguing in some cases that the magazine could be helpful in defusing industrial strife, only to discover that the majority of readers (or those engaged enough to respond to the survey), were married, middle-aged Protestant women. Some industries that had been taking the paper dropped it after running into union opposition or lack of reader interest.

Interpreting the data in the context of his industrial efforts, Peale reacted angrily, suggesting, ambiguously, that the paper itself was at fault. He complained to a staff member that the magazine lacked "punch," was "weak in its basic formula presentations," and had become "too soft and innocuous." What Peale obviously yearned for was what he understood as the robustness of evangelical Methodism, ordered up for competitive businessmen, instead of the material he was being presented, which, he explained, "I can hardly force myself to read . . . through and I expose it to myself on a reader-test basis."[40] LeSourd took responsibility for the "softened" image of *Guideposts*, and interpreting Peale's comments as a no-confidence vote in his editorial direction, offered to resign. When feelings had cooled for both of them, they examined their editorial differences again and agreed to some changes, although it remained clear that they were in disagreement about the fundamental question of the proper target audience.

An August 1968 story, written by a "typical" woman reader as opposed to a celebrity figure like Froman, illustrates some of the issues in dispute between Peale and LeSourd regarding questions of style, image, and target audience. Written in a way that is both more personal and intimate than Froman's account, it conveys a greater feeling of despair and frustration and speaks more obviously to an audience of women.

> Why did I stay with my husband through those years of disillusionment? . . .
>
> For Al came home [from the war], as did many soldiers, with a kind of haunted restlessness that seemed to be quieted only when he drank. And for Al, once he started drinking, there was no stopping point. Living with an alcoholic is indescribable to anyone who has not experienced it. The promises broken repeatedly, the shame and humiliation, the daily heartaches, etch little inroads of anguish into the soul.
>
> Over and over Al would promise to quit drinking, but each time he found a new excuse for not doing so. Over and over, humble and penitent, he would seek my forgiveness. And I would always give it, sincerely believing that next time would be different. . . .
>
> On one rare occasion when he attended church with me, he actually went forward to the altar and rededicated his life to God in sight of the whole congregation. This time I was certain the change was real and for days I was in heaven. Then he slipped back into the usual pattern and heaven crashed down around me. I continued my daily prayers for him, but they were hollow and heartless. . . .
>
> One day I read a small devotional article which drove home the thought that we are wrong to pray for the perfect situation in our lives. *The situation we are in*, the article said, *can be God's perfect place for us*. . . . Almost without knowing it I found myself on my knees, begging God's forgiveness in the first

> genuine prayer in many years. "And oh, Lord," I finished, "if this is where You want me to be, take my life and use it here, and give me strength to face each day."
>
> I shall never understand the events that followed this moment of truth. Something had changed inside me but I was not conscious of any outer change in my behavior toward Al. But within two weeks my husband stopped drinking, this time for good. Why a prayer for myself should so have affected someone else, I could not guess; I didn't question God, I just rejoiced.[41]

Other women who had managed to live productive lives with recovering or actively alcoholic husbands might have recognized in this account the author's ability finally to overcome denial and efforts to change the drinking partner, important insights in dysfunctional families of substance abusers. The help she needed apparently came to her through her own privatized religious devotions, as a result of despair, and not through the more traditional agencies of the church and helping professionals. Hers was not a topic likely to engage the attention of a male reading public, however, since all the evidence suggested that men married to alcoholic women abandoned their wives. The basic formula was essentially the same as in Froman's story, but here there were no war heroes, no dramatic rescues at sea—the kind of color that might add appeal for male readers. This was more a woman's story, about a woman once psychologically homeless or dispossessed finding new comfort and strength.

It was difficult for Peale to argue with the magazine's success. Subscriptions jumped from one million in 1961 to two million by 1973; the paper had so far outstripped its competition that it had no competition in its field. As the raw number increased, the proportion of industrial subscriptions decreased. It was still Peale's name that sold copies. Although there were a number of self-help publications similar in style, particularly from the Unity School of Christianity, it was not Unity's Charles Fillmore who wrote *The Power of Positive Thinking*, as a *Guideposts* executive noted with pleasure, but Norman Vincent Peale.[42] Yet Peale's personal experience with "muscular Christianity" and his disgust with "sissy religion" made it difficult for him to accept the fact that the audience produced by his message of harmonial Christianity and self-empowerment was predominantly middle-aged women. The men in his secular audiences who cheered his speeches remained largely unchurched and presumably essentially uninspired by the material in the new *Guideposts.*

During the winter of 1961–1962 relations between Peale and his editor became particularly stressful, due in part to their continuing differences over editorial philosophy, but more substantially to the fall-out from another Peale political venture. Peale had joined a group of conservative clergy who attempted to derail Senator John F. Kennedy's 1960 candidacy for the presidency, alleging that Kennedy's Catholicism raised issues of church-state relations. The short-range consequences of that event were significant for Peale both personally and professionally as reporters continued to dog his steps at the same time that newspaper publishers dropped his syndicated column and he withdrew his own name from speaking commitments.[43] He was sensitive, feisty, defensive. When LeSourd observed that perhaps this time it was his boss who might want to consider resigning, Peale reacted with uncharacteristic brusqueness at the suggestion of control. LeSourd apologized, but at the same time he spelled out his concerns: Peale, he said, had been reluctant to write *Guideposts* articles; he was "unwilling to attend" staff meetings; and he regularly gave the impression that most of the material in the magazine "isn't very good." It was an attitude, LeSourd added, that convinced him his boss felt himself "incompatible with *Guideposts*, its interfaith ideas, its large female readership, its editorial personnel, and its content."[44] The inside opinion at the paper was that, while the readership was not exclusively female, it was predominantly female. According to a senior staff member, as the paper's letters from readers grew from an annual trickle of 1,500 in 1949 to a flood of 500,000 by 1987, it was evident that the percentage of women letter writers remained constant, at about 90 percent.[45]

Although Peale might have preferred to read these demographic results in the light of the editorial policy at the magazine, his public devotional ministry, as contrasted with his appearances before business groups, had produced a largely female constituency. It was not the result he had anticipated, and he continued to deny it, pointing mistakenly to male majorities in the Marble Church membership, and relying heavily on male-based illustrations in his sermons. The Foundation for Christian Living, the organizing center for his ministry and the agency over which he maintained the closest personal control, tried to bridge the gender gap between his harmonial audience of women and his secular gatherings of men, although it was probably among the reading audience of his best-selling books that he came closest to acquiring a unisex following.

As if to signal his sympathy with Peale's concern for male readers, LeSourd himself wrote an article for the magazine in 1971 based on his interview with the country's most successful basketball coach at the time, John Wooden. Wooden had appeared in *Guideposts* before, and in this presentation LeSourd showed him as the virtual model for Peale's ideal "muscular Christian," the man who is tough, seemingly unsentimentally religious, and comfortable with the competitive aspect of secular society:

> Some UCLA players grumbled about [Wooden's] rules, according to [Lew] Alcindor, and felt that the coach was trying to make them prototypes of his own Horatio Alger, rural-America, Christian upbringing. During one game in the 1968–69 season, Coach Wooden had planned to insert into the lineup a player named Bill Sweek, but reasons of strategy changed his mind. Sweek, angry at being left out, walked off the court. After the game the coach lit into Sweek for his violation of team discipline. In the hot exchange which followed, the young man blurted out his suspicion: Wooden, he felt, gave preferential

treatment to players who conformed to his image of the wholesome All-American boy. . . .

Most expected Sweek to be dropped from the team. Wooden surprised them. "None of us is too old to change," he said. "I'm going to try to understand you better from now on." . . .

Since basketball today is a continuous 40-minute flow of running, jumping, sprawling players who can move fans—and coaches—into a frenzy of emotion, I asked Coach Wooden how he kept from becoming an ulcer-ridden insomniac. . . .

"From boyhood I've memorized great writings of all kinds," Coach Wooden said. "Doing this feeds my soul. I'll be driving along a freeway, upset about a practice session, or concerned about a player, when out of the depths will come some verse I memorized years ago. Repeating it lifts my spirit, and sometimes it will even give me an answer to what's troubling me."

"And that's how you keep calm during a game?"

"Yes, and—and with this." He took out his wallet and removed a small aluminum cross. "I often keep this in the palm of my hand during a game. It will help prevent me from jumping up and shouting at the referee. It also reminds me that basketball is not the most important thing."[46]

This rendering captured almost perfectly the sort of religious faith Peale hoped to generate among the businessmen who came to hear his talks. The accepted formula for a story—helplessness, death to the old self accompanied by a conversionary experience, and then rebirth to a changed life—was here altered to reveal a John Wooden consistently tough, if initially operating according to rules he recognized himself were dated and stiff. In other articles that told of emotional or physical healing, particularly if the subject was a woman, the outcome frequently focused on the person's new-found ability to accept hardships or through spiritual resources to transcend them. The Wooden story tellingly suggested mastery or victory over problems.

By the time LeSourd retired from the paper in 1974, subscriptions stood at two million, up from the 200,000 they had been when he became editor. Equally significant was the change in reader profile as "for-free" readers were replaced by those who chose the magazine for themselves. There was also a decisive shift in gender base. LeSourd had been an editor in good times and in bad: during the fifties when Peale's popularity crested in combination with the national revival of religion and then through the troubled sixties as his boss sought refuge from the Kennedy debacle and as the pressure of cultural events challenged all those engaged in public discourse. LeSourd's long tenure made it unlikely that his influence on the paper would be soon erased.

"The Lonely Man in the Motel Room"

In the first forty-five years of its history, *Guideposts* had but four editors, with Peale serving for most of that time as editor-in-chief and then as publisher.

Following Grace Oursler and Leonard LeSourd, Arthur Gordon served as editor between 1974 and 1981, when he was succeeded by Van Varner. Gordon was a close Peale friend, author of the 1958 authorized biography, *One Man's Way*, which became the basis for a Hollywood movie on the minister's life. He was also the first to bring extensive outside journalistic experience with him to the editor's desk. A self-described "closet humanist," Gordon trimmed some of the spiritual emphasis from the magazine and presented stories at once more stylish and contemporary.[47] Varner conceded that the paper had by his time become more secular, although "the faith dimension is still there," if somewhat more hidden.[48] With Gordon as editor, the journal increased in size while continuing to list extraordinary circulation gains: by 1982, the revised forty-eight-page monthly was being sent to four million subscribers, more than twice the number who subscribed to *The Christian Herald* and *The Christian Century* combined.

The paper took no advertising, was sold only through subscription, and by 1981—according to internal estimates—had reduced industrial subscriptions to a mere 8 percent of the total. Industrial subscribers were still solicited, but apparently for employers the attraction of *Guideposts* as a means of influencing workers' priorities had waned.

The glut of reader mail continued, almost all of it from women seeking help for themselves or others, and resembling in content and demographic profile the help seekers who contacted Peale through the other agencies of his ministry.[49] Varner reported that an article in the eighties on agoraphobia yielded 20,000 letters.[50] From its inception, the paper found that reader requests for help fell into three general areas. These were health, both physical and emotional; family relationships, including issues of divorce and child-rearing; and finance, especially how to find a job and how to pay the bills. The paper attempted to address these major categories while changing the presentation of the issues. The people who wrote in rarely suffered from the classically tragic ills of the world, as in the case of Job or Hester Prynne—although there were notable exceptions here—but were more like a Tennessee Williams' hero, caught in a crisis, temporarily overwhelmed by life's problems, with no resources readily available. Psychologically homeless, afflicted with a loss of history and continuity, they were withering from a narcissistic loss of a coherent self and feelings of self-esteem.

They may therefore have been just the perfect candidates for the form of self-healing the magazine proposed. Robert Fuller has suggested that alternate models of health care might actually be the preferred treatment for a particular category of patient, noting that "For many modern individuals, initiation into these metaphysical healing systems can thus quite literally open up new worlds and release inherent tendencies toward creativity, wisdom, and empathy for the needs of others."[51] *Guideposts* gave readers a simple formula for dealing with their problems, a context for interpreting their crisis, as well as entry into a surrogate "family." In the process of becoming more secular, the magazine redefined customary notions of what constituted the realm of the "spiritual." Even as the traditionalists in church and medicine continued to protest these new methods, the subscription list to *Guideposts* grew by leaps and bounds.

The paper continued to expand its services to the *Guideposts* family, producing international and Braille copies, sponsoring writing contests, offering church awards, and making available reduced-price copies of Peale's books. And it retained its tax-exempt status. It also demonstrated its popular appeal: By 1985, with four-and-a-half million paid subscriptions, it had twice the readership of a journal such as *U.S. News and World Report* and a million more than either *Newsweek* or *People* magazine. Peale's misgivings about its editorial philosophy and target audience had largely vanished. He had, in any case, located new audiences for his virile, motivational themes in the business and civic groups he regularly addressed.

Guideposts was a journalistic success almost in spite of Peale, and likely beyond even his generous expectations. It did not become the "organ for a great positive Christian movement" as he had hoped, but transformed from a political broadside of the Cold War, it achieved a more enduring cultural status along the lines of *The Reader's Digest* and *National Geographic*, the two periodicals with which it has been compared. No longer aggressively political after the late fifties, its editorial philosophy, inspirational and evangelical, generally conformed to a conservative civil religion. When a scholarly study surveyed *The Religious Press in America* in 1963, it concluded that such publications represented "largely an invisible phenomenon."[52] *Guideposts* was not even discussed by the authors, either because it was invisible to them or because it was not conventionally religious.

But the appeal of *Guideposts* is better understood when it is compared with its secular familiars, *The Reader's Digest* and *National Geographic*. The comparison works in the case of the former because of the clarity and crispness of its short articles and in the case of the latter because of its first-person accounts and cultivation of a family, or "Society," of readers. And all three can count on secondary readers in waiting rooms, hospitals, and even prisons. *Guideposts* is only selectively visible within high culture, in cosmopolitan circles, and among traditionalists in religion and medicine. But it succeeds where Peale himself succeeds, among "ordinary" middle-class Americans, particularly women. According to the paper's own evidence in 1981, the "typical" *Guideposts* reader resembled *The Reader's Digest* reader: *Guideposts* readers were "average in every way, except more religious,"[53] which could mean that some were "intolerant rednecks" on such matters as whether Mormonism is a religion or not. Average family income for a *Guideposts'* reader in 1981 was solidly middle class at $16,200, with other categories of sociological data remaining as they had been earlier: Over 80 percent of the readers were middle-aged (with an average age of fifty-three), middle-class women from Protestant backgrounds residing in an extended Bible Belt area reaching from Washington on the West Coast, across to South Dakota, then skirting Chicago before arcing through the South. South Dakota had achieved the highest density of subscribers, perhaps symbolic, speculated Editor Varner, of the isolation the paper had in mind when it targeted the "lonely man in the motel room."[54] A survey conducted by the magazine itself revealed that there were 3.6 readers per copy, encouraging the staff of the paper to claim a total readership in 1986 of 14,500,000, equally divided between men and women.[55] Internally generated data further indicated that 88 percent of *Guideposts* readers were churchgoers, with the divisional proportions among the three major faith groups remaining essentially as they had been in the past: The vast majority were Protestants, only 16 percent were Roman Catholics, and there were virtually no Jews.[56]

According to Arthur Gordon, as Peale began to devote himself more to other interests within his ministry, his *Guideposts* readers tended to lose track of his identification with the paper. For his part, Peale never lost track of *Guideposts*, but once its viability and growth were assured, he turned his energies to the personal challenge of new audiences. It became difficult to predict what the future of *Guideposts* would be once he retired from Marble Church in 1984 and devoted the bulk of his professional time to answering requests from business and commercial groups. By the first half of 1990, subscriptions dropped from a previous high of four-and-a-half million to 3,901,285.[57]

Notes

1. Although the Foundation for Christian Living was renamed the Peale Center for Christian Living in October 1991, I came to know it as FCL and have used the names interchangeably throughout.
2. Interview with *Guideposts*' editor Van Varner, New York City, June 18, 1987.
3. Peter Berger, Brigitte Berger, and Hansfried Kellner, *The Homeless Mind: Modernization and Consciousness* (New York: Vintage Books, 1974), pp. 83–96. Also Richard Quebedeaux, *By What Authority: The Rise of Personality Cults in American Christianity* (San Francisco, CA: Harper & Row, 1982), pp. 156–86; and Catherine Albanese, *America: Religions and Religion* (Belmont, CA: Wadsworth, 1981).
4. NVP to Dr. Dwight J. Bradley, National Citizens' Political Action Committee, December 27, 1944. Peale quoted from Bradley's letter to *Zion's Herald* in his response to him.
5. *Zion's Herald: Methodist Weekly*, January 17, 1945.
6. Ibid.
7. George Seldes to NVP, Sept. 14, 1945; NVP Ms. Coll.
8. Guy Emery Shipler to NVP, June 27, 1945; NVP Ms. Coll., Box 748.
9. NVP to John Sutherland Bonnell, Nov. 8, 1944; NVP Ms. Coll., Box 196.
10. NVP to Ralph Sockman, Dec. 21, 1944; NVP Ms. Coll., Box 196.
11. Bishop Herbert Welch to NVP, Dec. 26, 1944; NVP Ms. Coll., Box 748.

12. NVP to Harry Z. Harris, Oct. 9, 1945; NVP Ms. Coll., Box 743; Samuel J. Goldfarb to NVP, Jan. 10, 1945; NVP to Samuel J. Goldfarb, Jan. 29, 1945; NVP Ms. Coll.

13. NVP to George E. Stringfellow, July 21, 1947; NVP Ms. Coll.

14. NVP to E. F. Hutton, July 21, 1947, NVP Ms. Coll., Box 754; NVP to William Naden, Feb. 14, 1951; NVP Ms. Coll.

15. NVP to James D. Francis, Jan. 24, 1946; NVP Ms. Coll., Box 743. To the recipient of the letter, a member of the West Virginia Coal Board, Peale sent a copy of a clipping showing ministers walking a picket line, indicating that a copy of the clipping could be included in a financial appeal letter to businessmen.

16. NVP to W. N. Banks, Feb. 15, 1945; NVP Ms. Coll., Box 743.

17. NVP to Frank Gannett, Sept. 6, 1944; NVP Ms. Coll, Box 196.

18. NVP to Dr. Willard Johnson, National Conference of Christian and Jews, Jan. 30, 1945; NVP Ms. Coll.

19. NVP to Roy L. Smith, Editor, *The Christian Advocate*, Feb. 5, 1945; NVP Ms. Coll., Box 747.

20. Raymond Thornburg to Edward Rumely, May 28, 1946; Edward Rumely to Raymond Thornburg, May 29, 1946; NVP Ms. Coll. At one point, Rumely believed that Thornburg, Peale's associate at *Guideposts*, had contrived to gain access to the Committee's clergy mailing list without paying to rent it. They quickly settled their differences, agreeing that it resulted from a problem in communications.

21. NVP to Edward Rumely, Committee for Constitutional Government, Nov. 16, 1944; NVP Ms. Coll, Box 196. Peale's letter to Rumely asked him to add several names to the list for the *Guideposts* letter.

22. Memo of Edward Rumely to NVP, May 2, 1946; NVP Ms. Coll.

23. Edward A. Rumely to NVP, Nov. 20, 1944; NVP Ms. Coll., Box 196. Also E. A. Rumely to NVP and Raymond Thornburg, Nov. 2, 1944; E. A. Rumely to N. V. Peale and R. Thornburg, Dec. 1, 1944; NVP Ms. Coll., Box 196.

24. Alfred Haake to J. Howard Pew, Aug. 26, 1948; NVP to Alfred Haake, Sept. 21, 1948; NVP Ms. Coll. Peale thanked Haake for his reference to *Guideposts* in the letter to Pew and hoped that it might "bear fruit."

25. NVP, "Confidential Memorandum and Invitation," Nov. 1944; NVP Ms. Coll., Box 196. This document may have been edited before it went out; it is relevant here as an illustration of the early ideas that motivated the project.

26. NVP to George Stringfellow, July 21, 1947; NVP Ms. Coll.

27. Grace Oursler, "Editorial Report," Jan. 12, 1948; NVP Ms. Coll.

28. Ibid.

29. Grace Perkins Oursler to NVP, Mar. 8, 1949; NVP Ms. Coll., Box 756.

30. Raymond Thornburg to Stanley S. Kresge, Dec. 31, 1948; NVP to Honorable Brooks Hays, Nov. 19, 1949; NVP to E. F. Hutton, July 21, 1947; NVP Ms. Coll., Box 754. Peale's estimates were frequently somewhat inflated, especially if he was seeking contributions and wanted to describe a successful venture.

31. Raymond Thornburg to Stanley S. Kresge, Dec. 31, 1948; NVP Ms. Coll.

32. Memo from Len LeSourd to Mrs. Oursler, Dr. Peale, Mr. Thornburg, Mr. Decker, June 11, 1948; NVP Ms. Coll., Box 755.

33. *Guideposts*, April 1948.

34. In a letter to Peale from *Guideposts* managing director, Frederic Decker, he said that the lowest renewal rate was found on the "personal" list. Peale corresponded

with Decker on a regular basis to keep current with circulation figures. According to Decker, the low renewal rate was limited to the personal lists; "The figure there is 44%. As you will note, however, there is a very small number of subscriptions involved in these. Such lists as school teachers, etc., are classified as personal lists." Decker to NVP, Nov. 1, 1951; NVP to Frederic C. Decker, Dec. 3, 1951; NVP Ms. Coll.

35. NVP to Captain Eddie Rickenbacker, Dec. 21, 1961; Box 171; Leonard LeSourd to NVP, Mar. 29, 1962; NVP Ms. Coll., Box 321. This is the only early year for which I found evidence that might be pieced together to establish categories of subscribers. LeSourd said that 40 percent of the readers were "hard core," individual subscribers. Industrial sales had declined to 25 percent, and those who received the paper as a personal gift—usually as a Christmas present—represented another one third, about 33 percent. Another category, of 25,000, represented school teachers who were given the magazine by local contributors.

36. The term is one of those readily accepted and difficult to track down. A fairly common reference to the magazine, its original source is unknown.

37. Leonard LeSourd to NVP, Oct. 18, 1961; NVP Ms. Coll., Box 321.

38. *Guideposts*, April 1952, pp. 1, 2, 23.

39. Responses to *Guideposts Questionnaire*, Aug. 1957; NVP Ms. Coll., Box 285.

40. NVP to Starr West Jones, *Guideposts*, Dec. 16, 1959; NVP Ms. Coll., Box 286.

41. Jane Francis, "When Married to an Alcoholic," *Guideposts*, August 1968, pp. 12–15.

42. Interview with John Beach, Senior Projects Director, *Guideposts*, June 17, 1987, Pawling, New York.

43. See Chapter 7 for a fuller discussion of the Kennedy campaign.

44. Leonard LeSourd to NVP, Mar. 29, 1962; NVP Ms. Coll, Box 321. This letter was written in response to one by Peale, in which he was unusually sharp, admonishing LeSourd for suggesting he resign. Peale's letter also asked LeSourd to make sure his name was not connected in any way with a proposed *Guideposts* booklet on Communism, since he was still dealing with the consequences of the Kennedy affair. Peale said he wanted the magazine's staff to know "my determination never again to be involved in anything that might even directly be construed as political." NVP to Leonard LeSourd, Mar. 23, 1962; NVP Ms. Coll, Box 321.

45. Telephone interview with John Beach, June 26, 1987.

46. Leonard E. LeSourd, "The Elements of Victory," *Guideposts*, February 1971, pp. 12–15.

47. Interview with Arthur Gordon, February 24, 1981, Pawling, New York.

48. Interview with Van Varner, June 18, 1987, New York City.

49. The NVP Manuscript Collection is crammed with letters written to Peale in his various capacities: as author, minister of Marble Church, syndicated columnist, editor-in-chief of *Guideposts*, and as the head of both the Institutes for Religion and Health and the Foundation for Christian Living. Most are filed separately according to these categories. A random tally of letters done in seven-year cycles, from those addressed to him at FCL, noting the nature of the problem described and personal data on the letter writer—age, sex, location, religious affiliation—yielded results consistent with the *Guideposts'* profile. A major exception relates to the "Answer" column Peale wrote for *Look* magazine, which produced letters more representative of

the population as a whole; that is, the writers were more secular, urban, younger, more candid, and somewhat more evenly divided between men and women.

50. Interview with Van Varner, June 18, 1987.

51. Robert Fuller, *Alternate Medicine and American Religious Life* (New York: Oxford University Press, 1989), p. 136.

52. Martin E. Marty, John G. Deedy, Jr., David Wolf Silverman, and Robert Lekachman, *The Religious Press in America* (New York: Holt, Rinehart and Winston, 1963), pp. vii, 58, 59. The authors studied only the denominational press within the three major faith categories. *Guideposts* does not identify itself as religious, and it seems true that the authors used a traditional definition of religion.

53. Interview with *Guideposts* deputy publisher Wendell Forbes, February 24, 1981, Pawling, New York.

54. Interview with Van Varner, New York City, June 18, 1987.

55. *Guideposts* newsheet, provided by *Guideposts* special projects director John Beach.

56. Interview with Wendell Forbes, Pawling, New York, February 24, 1981.

57. Telephone conversation with Terry Panny, secretary to *Guideposts* president John Temple, September 25, 1991.

CHAPTER 5

Positive Thinking and the Cure of Souls

My next point is that in the scale of powers it is not talent but sensibility which is best; talent confines, but the central life puts us in relation to all.

. . . Aristotle or Bacon or Kant propound some maxim which is the key-note of philosophy thenceforward. But I am more interested to know that when at last they have furled out their grand work, it is only some familiar experience of every man in the street. If it be not, it will never be heard of again.

. . . One more trait of true success. The good mind chooses what is positive, what is advancing—embraces the affirmative.

Emerson
Essay on Success
1858

By the 1950s, Peale had become a celebrity of national, probably international, proportions. When a social critic attempted to find an explanation for the phenomenal response to Peale's message during the decade, he wrote in *The Saturday Review*, "Where there is a hunger there must come a feeding."[1] Seemingly more stymied than other critics who tried to explain the correlation between the religious revival in progress and Peale's enormous popularity, he dismissed Peale's stunning success as "Couéism with organ music."[2] Ersatz food, he believed, was feeding America's spiritual hunger. Other commentators on popular culture, however, were inclined to view the revival of religion somewhat more favorably. In 1955, as the statistical evidence on the movement crested, the New York *Sunday News* said approvingly that Peale, Billy Graham, and Bishop Fulton J. Sheen, were the leading contributors to a major religious development, which some scholars have come to call a "Fourth Great Awakening."[3]

The force that catapulted Peale to the top was his 1952 book, *The Power of Positive Thinking*, and the overwhelming reception it received. The unpretentious volume, which appeared in bookstores in October 1952, just a month before Dwight D. Eisenhower's landslide victory over Adlai E.

Stevenson, changed the direction of Peale's career and had an important impact on the course of American religion. When asked to summarize the theme of the book, Peale described it as a "point of view," that "good overbalances evil." In his estimation positive thinking was "an individual attitude," not a "public opinion" position. He said it had a biblical source, best encapsulated in Jesus' statement, "Be of good cheer. I have overcome the world."[4]

In many ways, the popular success in the 1950s of a book like *The Power of Positive Thinking* was predictable. It had obvious reader appeal in its unremitting emphasis on self-empowerment and peace of mind. Moreover, it appeared at a time when religion was "in the air," when religious popularizers and media specialists cooperated on "new methods" of evangelism, when Hollywood star Jane Russell thought it timely and appropriate to refer to God as "a livin' doll." Few segments of society were unaffected by the wave of interest. Serious theologians like Reinhold Niebuhr and Paul Tillich became celebrities of a different sort to intellectuals and other participants in high culture. Peale, Billy Graham, Fulton Sheen, and Rabbi Joshua Liebman, the latter the author of *Peace of Mind*, produced material more accessible to middle-class audiences, while the press reported the record numbers of people, presumably largely of working-class backgrounds, that descended on Yankee Stadium to join with Jehovah's Witnesses.[5] Yet it was Peale's message, revealed simply and starkly in the book, that gave definition to the religious revival. Along with Billy Graham's evangelism, it helped set the pattern for trends that would develop within American religion over the next two decades. Although its popular appeal was obvious, it was nevertheless remarkable that so unprepossessing a book could captivate and define a movement.

There were social observers who looked to sources in the culture to explain the reasons for the revival itself. They noted fears of the Cold War and atomic bombs, of problems of economic insecurity and "other-directed" role expectations, of the emergence of popular psychology, as well as the skill of clergy leaders in marketing religion. But it was impossible to deny the impact of Peale's singular message. Even his most caustic critics admitted at the time that fundamentally the Phenomenon of Pealeism could be traced to the public response to a simple concept: "It is an idea that has made Dr. Peale," noted the often quoted Peale detractor William Lee Miller.[6] According to Miller, while it was Billy Graham's personality and his "virtuosity as a performer" that explained his success, it was more Peale's message, his steady reiteration of the simple concept that "affirmative attitudes help make their own affirmations come true" that accounted for the New York clergyman's vast following.[7] Incarnated in *The Power of Positive Thinking*, that message and the book became the center of a national debate over the appropriate role of religion, a storm that subjected Peale to what his authorized biographer called "The Rage of the Intellectuals."[8]

Peale's national audience and signs of growing religious interest had both been building at least since World War II. His first self-help book, one he called a "formula book," was *A Guide to Confident Living*, put out by his new publisher Prentice-Hall in 1949. It capitalized on the prevailing religious mood, becoming a best seller and moving the Peales into a new tax bracket. Statistical evidence for a "turn toward religion," in terms of religious affiliation, had started accumulating as early as the turn of the century, until by 1950 the data measuring the proportion of "church members" to the general population showed that over half the population, or 57 percent, claimed religious affiliation, a number that by 1956 reached 62 percent.[9] There was, however, a seemingly paradoxical correlate which revealed that this amazing growth coexisted with declining commitment to traditional religious orthodoxies. Predictably, the evidence also made clear that the religious revitalization effort was sustained by the evangelical Protestant middle class, who were shaping it into a populist manifestation of their special interests and preferences.

Consistent with the way anthropologists have defined revitalization movements, the religious resurgence in the fifties can be explained as a grass-roots effort aimed at reclaiming a kind of tribal past; it defended a unique form of popular culture that to adherents appeared threatened by the values of a traditional religious bureaucracy.[10] In combination with this defense of a distinctive past, there was an assertion of entitlement, essentially to the varied services and resources of the present. For such a movement to succeed required spokespersons such as Graham and Peale—who held different places on the theological spectrum—as faithful reflections of those values, to synthesize these views of a distinctive past with a real present in a way that appeared true to the value structure of the group.

As his national audience mushroomed, Peale claimed that its numbers were provided largely by those previously unchurched. That appears not to have been the case, but rather, as with his *Guideposts* subscribers, his growing audience of enrolled supporters consisted largely of mainstream evangelical Protestants.[11] Their support for him during the 1950s did not affect their continuing membership in their local congregations. Instead, the national movement made the reality of dual religious citizenship—supporting Peale's positive thinking network while still keeping membership in mainstream churches—more common, giving rise to positive thinking interest groups that transcended denominational lines. In time, these divided loyalties would weaken both ecumenism and the authority of mainstream organizations.

Repeatedly savaged for his popular views during the decade by theological critics, Peale began to withhold his support for the kind of generic Protestantism he once identified with "the Militant Church." He invested most of those former institutional loyalties in his own organization, the Foundation for Christian Living. The Foundation, from its Pawling location, became the clearinghouse for all Peale activities and the center of what by mid-decade could only be called an empire. Positive thinking, as practical technique as well as the byword for the religious revival, helped alter traditional middle-class notions about the role of the church in matters of personal life, such as health and life crises, and contributed to a growing cultural emphasis on individual self-governance. Given what was at stake, the liberal orientation and institutional power of the mainstream leadership, it is not surprising that Peale and his critics often seemed locked in deadly verbal combat. But if Pealeism had the negative effect of helping initiate a trend that would diminish the influence of mainstream churches, it had the positive cultural consequence of causing those churches to be more self-critical and more responsive to the needs of their local members. It managed to accomplish what R. Laurence Moore has observed to be a byproduct of American dissenting movements generally—namely, their ability to expose "the shabbiness and the arrogance of the culture surrounding them," thereby contributing "a fair measure to whatever success the American system has had."[12]

"A Seedbed Source"

The overwhelming reception accorded *The Power of Positive Thinking* soon made Peale a "minister to millions."[13] By his own estimate, he reached a weekly audience of 30 million people by 1957. The media, quickly discovering that detailing the frantic schedule of the "jet propulsion man" made good copy, provided steady updates on his accomplishments. In a lengthy and careful article on the revival for *Life* magazine, *Christian Century* editor Paul Hutchinson observed that Peale "preaches to probably the largest audience ever gathered by an American cleric."[14] The estimates were impressive: Two Sunday sermons at Marble Church reached 4,000 people; his syndicated newspaper column reportedly attracted 10 million weekly readers; his weekly radio program produced 3 million listeners; *Guideposts* magazine claimed 500,000 subscribers; his three published sermons monthly went to a world-wide audience of 150,000 people; and there was *The Power of Positive Thinking*, which in two-and-a-half years on the best-seller list sold over a million copies. Over the next forty years it would continue to be a trend setter in sales. There were myriad other Peale projects, including an advice column for *Look* magazine, a Mr. and Mrs. television show with Ruth Peale, a recording of the book as well as a youth edition and a pocket edition of it, regular contributions to *The Reader's Digest* and other popular periodicals, extensive speaking engagements to business and civic groups, and a line of Hallmark greeting cards. It was easy for the media to refer to him as an industry and for the public to perceive him as a celebrity, the first modern clergyman to be so lavishly feted by publicity.

It was evident that Peale's popular reception and the religious revitalization effort were related, although there was no obvious cause-and-effect relationship. What they shared was an historical context: The two developments coincided at a critical historical moment, a time corresponding to what Talcott Parsons called a "seedbed source" of significant cultural change.[15] It was a time, too, when Robert Wuthnow says American religion was in the process of being "restructured." Pealeism was a major factor in the process of religious and cultural restructuring. Especially in the years between 1953 and 1955, when *The Power of Positive Thinking* was on the best-seller list and Peale not only became a household item—showing up on quiz shows, in crossword puzzles, and as the assumed name of a striptease artist—but was the subject of intense critical debate, a number of important cultural developments converged that helped establish a religious agenda for the future.

The success of Peale's public offerings and Graham's fervent revivals tended to polarize the religious community, less along class lines than along interest lines.[16] Although it was relatively simple to posit a conservative-liberal cleavage, the separation into interest-group categories was more complicated than that. Divisive Cold War political loyalties had their counterpart in sympathy for either the liberal National Council of Churches or the conservative National Association of Evangelicals, the latter clearly reflective of the preferences of Graham and Peale. And then there were the ways that personal religion, from positive thinking clubs to subscriptions to *Guideposts*, differed from the challenge of social Christianity. Furthermore, Peale's rough handling by his critics, which included charges of blasphemy, heresy, and occultism, prompted him to consider leaving the mainstream church. That he did not actually leave was due, he said, to an unspoken promise to his father, who died in 1955. But in most ways that mattered, he did in fact leave, for although he continued to respect his nominal commitments to denominational responsibilities, his considerable energy and emotional attention were thereafter directed to the Foundation for Christian Living, the hub of his independent ministry. The success of the book, coinciding with his popular appeal during the revival and his harsh treatment by the liberal critics, clearly empowered him to focus his interests, to develop his independent ministry through the Foundation for Christian Living. It was an example not lost on his supporters.

The times generally were conductive to stirrings of revival and revitalization. The picture of postwar America etched by historians, economists, and other observers of the era creates the impression of a scene of endless motion, where evidences of expansion and pent-up demand mask incipient signs of cultural conflict. People were literally rootless as they moved from suburb to suburb and job to job and psychologically and spiritually disconnected in a period shaped by atomic anxiety and ethical uncertainty. Signs of religious growth were one response to this dislocation, with church membership and attendance figures soaring as church building projects proliferated to keep up with the unprecedented expansion. But linked to growth in the institutional church was the troubling and unexplainable evidence of decline in traditional beliefs.[17] Many theories have attempted to explain the outward turn to religion, and they refer to such obvious cultural markers as the pietistic aura of the Eisenhower administration, which added the words "under God" to the national pledge; the challenge of Communism; the social rewards of church membership; the emptiness of materialism as an ideology; and the new expectations for personal and family enrichment.[18] Sermons of popular preachers encouraged listeners to cultivate a dynamic, vibrant faith in keeping with the expansionary mood of "a time for new beginnings."[19] Quickly, very quickly, however, this burst in religious expression began to deflate, with 1958 the high point for religious attendance, so that by 1959, as Sydney Ahlstrom noted, "discerning observers began talking about the postwar revival in the past tense."[20] By 1960 signs suggested the religious revival was ending, a premature demise, Peale believed, hastened by the carping of liberal critics.[21] The darker side of the social statistics was coming to light.

Retrospectively it is possible to see the major paradox of the movement, the decline in religious orthodoxy and the expanding church memberships, as a measure of other contradictions within the culture, and as a means of hedging one's bets. The era offered promise, but it also suggested peril; its expectant optimism coexisted with uneasy pessimism; and signs of fear vied with displays of faith.[22] The hope inherent in geographic and job moves competed with a fear of personal inability to cope in a more competitive environment. In voices high and tight, people asked what would happen, to them and to the nation, if there was another depression or if the Korean War widened or if they lost their jobs and couldn't make the mortgage payments. The creedal religion of the traditional churches, with their social programs and limited personal ministry to individuals, no longer seemed enough. Peale listened and suggested remedies. When the revival started to founder, he contended that liberal charges about its superficial nature hastened its demise, although it was more likely related to these broader cultural contradictions.

Then, too, Pealeism made it possible to claim to be "religious" without conforming to old standards of orthodoxy. Orthodoxy, Peale could argue, did not decline at all during the decade, it was simply redefined by new sources and standards. Added to that was the autonomous noncommunal nature of the very movement Peale himself had sparked. There was, finally, the fact that the so-called third generation of Americans to which Will Herberg referred was passing its prime as social arbiters, willing its legacy of values and authority to its children, the yet more vocal opponents of establishment authority in the sixties.[23]

"A New Thought Classic"

A kindly disposed reviewer of *The Power of Positive Thinking* described it as "wise, friendly, unpretentious guidance . . . about the widespread . . . misery of a nation rich in things, but poor in spirit."[24] While Peale's detractors associated it with a host of darker themes, J. Gordon Melton, analyzing it some twenty-five years later, more accurately placed it within its appropriate historical context, as "a New Thought classic."[25] Simply written, easy to understand, the book employed language and traditional references—to church going, prayer, Bible reading—that anyone from within America's mainstream religious tradition would readily comprehend. But because the New Thought tradition frequently assigned familiar religious words new, specialized, and variant meanings, it was easy for a reader unfamiliar with the nuanced implication of terms such as *Mind*, or *supply*, or *law* to miss their New Thought context.

Peale had been examining the literature of New Thought and the mental science tradition since about 1928, but metaphysical positive thinking did not become the distinctive part of his public message until about 1949. Given his particular history with Methodism, it was a relatively shorter reach for him spiritually to New Thought than to Calvinism. It was a leap of faith replicated by many other continuing and former Methodists, who, along with former Universalists, found the message of New Thought particularly congenial and were well represented in its local groups. Peale evaded direct public connection with New Thought and the mental science experience, insisting that he was a "preacher of the old-fashioned gospel": When an article of his appeared in the *Science of Mind* journal, he observed, "They think I am one of them, but I am not. Actually I am not opposed to it."[26]

Peale's practical Christianity was in fact a euphemism for his unique version of New Thought, politicized and evangelicalized. New Thought was that branch of the harmonial metaphysical tradition which traced back to Ralph Waldo Trine, Charles Fillmore, Julius and Annetta Dresser, and even the pioneering Phineas Quimby. As New Thought began to grow at the turn of the century, it wisely maintained its schedule of Sunday afternoon meetings, allowing members so inclined to attend more traditional church services in the morning, thereby creating the opportunity for dual religious citizenship. His book was, however, more than a restatement of New Thought principles and tradition: It was a pragmatic application of those ideas to the contemporary scene.

Peale was able to respond to shifts in the culture like a jockey on a thoroughbred, comfortable with adaptations to change for reasons part psychological and part ideological. Never convinced by the theological critiques about the moral compromises inherent in a Christ-in-culture model, he regarded his ministry as a therapeutic one, of necessity driven by the needs of members as either patients or clients. His work was a healing one, he believed, as in the ancient cure of souls, claiming that "This is a return to the original practice of Christianity."[27]

The book appeared at a time when Peale's ministry itself was changing, taking on a different emphasis, which may help account for the fullness of its acceptance of New Thought concepts. His earlier shift from the Methodist Church to the Reformed Church in America had dimmed his enthusiasm for denominational ties generally, but he kept a nineteenth-century Protestant faith in the power of "The Church" and its ability to effect social change. It was a commitment that carried over into the sort of active politicking that he did during the 1940s. But hammered by liberals both inside and outside the church for his political views, and recognizing that his partisan activities adversely affected his church-based ministry and the growth of *Guideposts*, he withdrew by 1950 from his earlier level of political involvement. If liberals were going to challenge him for both his politics and, less directly, his ministry, then he would find new avenues for expressing these related concerns, outside the church if necessary. He adopted a lower political profile, working quietly for a few years with an East Coast variant of Spiritual Mobilization, and then for a time flirting with H. L. Hunt's Facts Forum. And gradually, for reasons pragmatic and ideological, Peale reoriented his ministry.

At the time Peale was introduced to New Thought in the late 1920s, it had been a viable, marginal, religious option in America for nearly a half century. Its partisans traced its modern roots to the Maine healer Phineas Quimby, who in the mid-nineteenth century began treating patients through a form of suggestion that relied on the mental (or spiritual) ability of the healer and client to tap into the restorative power of divine energy. Some of Quimby's patients—Mary Baker Eddy, the Dressers, and Warren Felt Evans—soon became better known than their teacher. They began teaching his methods while simultaneously creating the nucleus of a religious movement. Factionalism, based on personality issues and differing interpretive perspectives, caused the movement to divide into two wings: Christian Science, which endorsed a radical idealism in its emphasis on the role of Mind alone, and New Thought, which called itself the "liberal" wing of the movement.

Outsiders came to regard New Thought less as a religion than as a secular movement concerned with healing and personal well-being. They chided it for an assumed monism, for neglecting Christ's atonement, and for naiveté about the reality of sin in the world. The whole mental science movement with which New Thought was allied tended to be dismissed by the theologically learned as the parlor pursuit of middle-aged women. And it was quite true that not only did women predominate in New Thought, as they did in other Protestant groups, but they were frequently found in the organization as teachers, healers, and leaders. New Thought came to be included in what Union Theological Seminary professor John C. Bennett called "culture religion," or what others saw as a sectarian model of America's civil religion.

Peale brought to his interpretation of New Thought a history of Methodist evangelicalism, Calvinist language, and conservative political activism. Eclectic, synthetic, obviously uncreedal and unsystematic, his particular theological blend combined aspects of evangelical Protestantism, metaphysical spirituality, and the American Dream; it proved to be perfectly adapted to the cultural currents of the decade. Activated by the dramatic sales of *The Power of Positive Thinking* and Peale's role in the revival, this popularly appealing belief structure shifted from a marginal cultural position to the mainstream, eventually adapted by both liberal and evangelical religious groups in counseling, and absorbed within a conservative civil religion.

Researching a Popular Success

It had taken Peale four years to extract writing time from his hectic schedule to produce what would become his most famous book. He finally committed to the writing during the summer of 1951 while on a trip to Hawaii sponsored by Spiritual Mobilization, turning the occasion into a working holiday with the assistance of Ruth, a typewriter, and a dictaphone. He had been urged into the project by his new editors at Prentice-Hall, who continued to calculate record sales for his previous work, *A Guide to Confident Living*, reportedly still selling 1,000 copies weekly.

The financial windfall from book royalties helped contribute to the change in his ministry. Book sales made the family independently wealthy, which meant that while old habits of frugality still prevailed—Ruth still preserved jams and jellies at the farm and stored sale-priced turkeys along with their own farm-raised beef in the freezer—Norman could begin devoting more of the monies he raised on his speaking swings through the country to his various ventures. It meant, in addition, that he could enjoy more freely his special pleasure of trying out the latest consumer gadget and following his passion for luxury cars, more legacies of an impoverished youth. An adventurous consumer, he was a salesman's delight. In terms of his ministry, however, book royalties from *Confident Living*, but particularly *The Power of Positive Thinking*, granted him ecclesiastical and denominational freedom and the opportunity to channel his new-found resources into the development of the Foundation for Christian Living. He was relieved of the endless embarrassment of appeals for money from his national audience of readers and could support his projects with the income from royalties, speaking engagements, and the generous gifts of wealthy benefactors. Nevertheless, while *Confident Living* sold well, it would eventually seem less accessible than its successor.

In time the author of forty-three books, Peale once confessed, "I am not, never was, nor never will be a writer."[28] Writing was, of the many activities that engaged him, the one he claimed to like least because it was "hard work." Yet he soon came to appreciate his books, not just for the financial freedom they produced but for their ability to have his diverse constituency meet on common ground. His mission, he stated, was to reach the unchurched, and surely his books and newspaper columns provided the best line of access to them. His books, but particularly *The Power of Positive*

Thinking during the 1950s, enabled his diverse audience—collections of businessmen, middle-aged women, and the unchurched—for a time at least to become a congregation.

The celebrated book did reveal the qualities the kindly disposed reviewer noted: It was friendly, unpretentious, and concerned about the misery caused by spiritual poverty. In its ebullient, upbeat tone, it belonged more to the success genre of New Thought literature than the inspirational variety, although there was a good deal of evidence of the latter. Essentially a collection of anecdotes, it had an overwhelmingly male bias in both content and cast of characters. In some ways, it read more like a doctor's log after grand rounds through a men's ward than the journal of the itinerant preacher Peale had become. The writer seemed to be addressing people much like himself, the mythical archetypal reader of *Guideposts*, the "lonely man in the motel room." In his endless odysseys across the country he had run into countless people, usually men, for whom the demands of the ordinary, of the daily pressures of work, family, and society, had obscured a sense of personal meaning and their health was suffering. *The Power of Positive Thinking* told their stories and the remedies Peale prescribed for their ills. The book was therefore premised on two modest assumptions: that there were significant numbers of people who thought they were sick, in need of care, and incapable of being fully productive contributors to society; and that they were sufficiently interested in being restored that they were willing to try his therapeutic message of practical Christianity, and not incidentally, buy the book.

It appeared as if Peale had collected the raw material for his anecdotal exposé of the maladies of middle-class businessmen over the years in his wide-ranging travels. He also managed to report on a few entrepreneurs and some wives. For a person always on the move, serious reflection was clearly out of the question, as the book showed, but the very thinness of the analysis allowed the stories to spill out in a kind of stream of consciousness. His confessors were traveling salesmen, conventioneers, hopeful deal makers; temporarily homeless, they revealed evidence of psychological and cultural rootlessness. They complained of fatigue, of "cracking up" and breaking down, of insomnia, of heavy drinking, and of loss of energy; of difficulties that in many ways seemed milder than those described in the many letters that arrived daily at his office written by women. The physical and psychic crises of everyday life that in an earlier time were tolerated, medicated, and nurtured away had become costly economic liabilities in the new competitive milieu. But it is probably also true that middle-class people had come to expect "more," to anticipate a better quality of life. It was part of the promise and the peril of the times: If the new environment provided less personal security and support and more external control, it also suggested the possibility of greater rewards.

The book's reception made Peale a public personality; its commercialization made him a celebrity. Although linked with the newer reassuring "psychological" works such as Joshua Liebman's *Peace of Mind*, Fulton J. Sheen's *Peace of Soul*, and later in 1953, Billy Graham's *Peace with God*, it remained in a class by itself. Even after leaving the best-seller list, the book continued to sell at an astonishing rate. It went through repeated printings, was reproduced on records, tapes, and cassettes, and then made into a series of films, so that by the late 1980s it had sold over fifteen million copies. At that late date it still remained among the top ten self-improvement books, and according to one of Peale's associates, had become "the bestselling non-fiction book by a single author in the history of publishing." Members of his staff were predicting an entirely new audience for it among young people just entering the book-buying years.

Why did the book sell? To say that it was because Peale was popular and the book was carefully marketed is too simple, though both conclusions were correct at the time the book appeared. People bought the book because, as they said repeatedly in letters to him, they felt they were leading shriveled, diminished lives. Peale offered them a remedy, which like a physician's, was less immediately concerned with the cause of the malady than its treatment. Inevitably Peale located both the source and the cure for the problem in the individual, not in the society.

Creating a Best Seller

Peale never intended *The Power of Positive Thinking* to be a theological discussion, nor to be subjected to serious theological analysis. He apologized for its limited literary qualities and said it was designed to be a "practical, direct-action, personal-improvement manual."[29] Given his own religious credentials, however, and the basic premise on which the work was based, namely, the belief that "God is in you," he was inevitably drawn into creating a religious/spiritual context for his approach to healing and self-help.

The three theological influences on his thinking are evident in the work—that is, religious liberalism, evangelicalism, and the ideas of metaphysical New Thought. The book is an optimistic, post millennialistic, immanentistic statement about the reality of the realizable wish: With positive thinking and prayer, Peale contended that "You do not need to be defeated by anything."

The central symbol in the book is the metaphysical concept of Mind. Its role was to engage the senses and the will, and to respond to the rational mind. In Peale's usage, Mind, or what had been for the English researcher Frederick W. H. Myers the concept of a positive subliminal self, became synonymous with soul or the subconscious and therefore held none of the dark repressions attributed to it by Freud. But significantly, like Freud's view of the libido, Peale's argument described this mental force as powerful, generative, and dynamic, as many later New Age groups would as well. Activated and informed, or "filled" by the senses, the will, and conscious mind, Mind became a repository of powerful healing and creative forces. For that power to be released, however, it was necessary for the conscious mind to surrender and to be absorbed within the Divine Presence or Mind. In the language of the text, it must become a channel and participate in the divine flow of Energy. Because Peale's theology—and his imagery—spanned two traditions, that is, his Methodist/Calvinist background and his New Thought adoption, his language and symbolism were never firmly anchored in the metaphysical mode.

The controlling imagery, however, derived from the metaphysical tradition of New Thought: Mind as channel for divine Energy flowing through the cosmos.[30] It was an imagery that was naturalistic, holistic, primal, and gendered. It synthesized two beliefs held by Peale and developed in the work, namely, commitment to holistic, psychosomatic medicine and to a concept of personal growth that extended to personal immortality. But there was also some, if limited, evidence in the book, as in his sermons, of a more orthodox alternate imagery, which conveyed less a sense of unity and cohesion than of separation and distinction. It reflected a dualism consistent with mainstream Protestantism and the technology of modern culture. In this latter context, the imagery was more assertive and contingent: God's energy was not harmonious but atomic; God's will a precondition for all believers; sin not unknown. Essentially, however, the work argued that religious belief needed to be pragmatic, in tune with the culture, and verifiable by certain "scientific" laws or principles. Missing in the discussion was any explicit incarnational commitment or acceptance of the enduring reality of sin in the world. Sin, like sickness, was assumed to be transitory and created by blocked channels of energy resulting from fear, resentment, or poor parental influences.

The book was structured as a collection of stories, told in the first and third person, about people's problems and the psychospiritual remedies Peale recommended for them. The therapeutics, because their source was said to be "scientific laws," were defined as repeatable and predictable, available to anyone properly prepared through thought conditioning to make use of them. The technique common to all efforts to produce change was positive thinking, or mental suggestion, and it was based on the three-step approach Peale had been advancing for some time: to picturize, prayerize, and actualize.. A person was advised to "picturize" a desired goal, such as restored health or a revitalized relationship. With the image in mind, he or she was to engage in affirmative prayer to learn God's will about the objective. If it was determined that God's will supported the goal—and fervent affirmative prayer was apt to make that likely though not inevitable—then the person was expected to take practical steps to bring it about.

Concluding chapters in the book discussed Peale's views on death and immortality and showed his indebtedness to New Thought on the subject. Recounting stories of the dying and those who had near-death experiences, he summed up with the metaphysical affirmation, "there is no death." He had long been attracted by theories of parapsychology, believing himself to be possessed of special gifts in the area, including the gift of clairvoyance. He speculated that the time was not far distant when science would substantiate some of the theories of parapsychologists, such as "the existence of the soul and its deathlessness." That had been the similar desire of narcissists through the ages.

Though the book was broadly and popularly written by design, some parts strained credibility for modern secular readers. The skeptical surely wondered at the prospect of busy executives praying before an important meeting or sharing their methods for daily Bible reading over hotel breakfasts. For those untrained in psychotherapy it was difficult to know which of his behavioral modification techniques might be effective and which were useless, if not silly. Peale recommended throwing your calendar in the trash at the end of the day; concentrating on emptying your mind; asking "please" when soliciting God's help; remembering people's names. In terms of behavioral change, however, the book contained sound advice on dealing with alcoholism and in its references to the Higher Power revealed Peale's past friendship with Bill Wilson and his work with Alcoholics Anonymous.

Many of the book's themes were familiar to liberal theologians and to physicians of psychosomatic medicine. Predictably, the book has enjoyed greater support over the years from the medical community than it has from theologians and members of the clergy. The clinician is first concerned with the restoration of health and the instrumental value of techniques. Theologians contend that religion was never meant to have instrumental value, although on occasion it might fill that ancillary role. Authentic religion, they argue, must be focused on eternal truths and ultimate values. Although Peale would note in his defense that a close reading of the book would demonstrate that the true goal of positive thinking was indeed focused on an eternal value, namely, a redeemed life lived in close proximity to God, this was not the conclusion most would readily draw from the work. The popularized presentation, in both writing and content, made the book available to a large public, but it did so by reducing the discussion to simple, even simplistic terms, and by making pale images of more complicated concepts.

The Populist and the Critics

Few books in modern times have been greeted by a more polarized response, an important commentary on both the work and the culture. The enthusiasm of the popular reception was matched by a savage critical attack, which appeared to occur in choreographed wavelike motions over the decade. Critics in seminaries, universities, and mainline churches were legion. Sophisticated secular publications, like *The Saturday Review*, predictably attacked it, but so did *Christian Century* as well as *Time*. Executives in church agencies and denominational officials joined in the onslaught, motivated by genuine intellectual conviction, though undoubtedly mindful that Peale's institutional critiques undermined the ecclesiastical infrastructure they themselves represented.

When Ralph Waldo Emerson faced a similar chill wind for his views, he referred to its source as that "ice house" of formal religion, Harvard Divinity School. Peale found the counterpart for Emerson's nemesis in his near-neighbor, Union Theological Seminary, Union's faculty, and most of its students lined up squarely against him. Reinhold Neibuhr admitted that he and Peale agreed on absolutely nothing. Dean Liston Pope of Yale Divinity School criticized the development of a "cult." Bishop G. Bromley Oxnam of the Methodist Church chided the author for his theology as well as his lack of social awareness.[31] Particularly biting articles by William Lee Miller in *The Reporter* and William Peters in *Redbook* had devastating effects on Peale.[32] His close associates and loyal family, particularly Ruth, aware of his ready vulnerability to slight, had tried over the years with considerable success to shield him from criticism. As a result, he was unprepared to confront this serious and sustained attack on his work.

The ferocity of the struggle between Peale and his opponents was an indication of the magnitude of what was at stake for religion and the culture. Peale, of course, was not without his public defenders, and he and they had a forum for their views in journals such as *The Reader's Digest, The Christian Herald*, and *Christianity Today*. Among the weekly newsmagazines, *Newsweek* tended to be kindlier in its reports of Peale than *Time*, although both were capable of mixing praise and criticism.

The harshest criticisms came from the journals reflective of high culture: *The Saturday Review, The New Republic, The Nation, The Atlantic Monthly*. The criticism tended to divide into three categories: one was directed toward Peale personally; another viewed the cultural consequences of the movement; and another took on the theological or intellectual issues. The personal comments usually consisted of innocuous name calling, the cultural associated Pealeism with the tawdriness of popular culture, and the intellectual identified his published views with historically dangerous and insidious ideas. Of the three, it was the last, which effectively named Peale a heretic, that was most painful to him and to Ruth. Almost thirty years after the verbal storm, the recollection of its fury could still bring Ruth to tears.[33]

The name calling reached for the obvious and was largely inoffensive. Since the bestselling book so clearly packaged positive thinking as a product, so frequently invoked the techniques of advertising, and so unashamedly commercialized its appeal, it was to be expected that critics would connect Peale to the title of a 1949 article about him, "God's Salesman." References to salesmanship and advertising abounded in the critiques: Peale turned the table on his attackers and wore the badge of religious salesman with pride. It worked to his advantage in his innumerable contacts with salespeople at their annual meetings, implicitly suggesting a "we" versus "they" bonding. Other labels were even thinner. William Lee Miller referred to him as the "rich man's Billy Graham," while another reviewer wrote an article about him under the title "The Confidence Man"; another said he offered "pantheism in suede shoes and a gray flannel suit," decidedly incorrect if meant to characterize Peale's extremely conservative attire.[34]

The cultural comments, many seemingly intended to be only explanatory, were fundamentally the most relevant and consequential of the attacks, for they depicted two cultural styles, even two cultures, separated by a wide chasm. It was as if some representatives of high culture regarded the consumers of popular culture as members of a different nation. On one side were the critics, distant, observant, authoritative. On the other was "the Gospel Boogie," "Hollywood and Tin Pan Alley," and "The Cult of Happy Living."[35] Critics who took pains to detail the evidence of popular culture's tribute to religion inevitably scorned or trivialized the product. They referred derisively to the government's addition of "under God" to the official pledge and its plan to include "In God We Trust" on postage stamps. They criticized the vapid religion evident in movies like *The Robe* and *Salome*. And they exposed the simplistic theology contained in popular songs like "I Believe," "Have You Talked to the Man Upstairs?" and "It Is No Secret What God Can Do." Donald Meyer was one of the few commentators to recognize that advertising and technology had transformed popular culture, or what he termed "folk culture," over the preceding thirty years. Yet even he was led to conclude that the search for "new standards of happiness" had led instead to "status panic and status apathy" and to feelings of guilt and inferiority.[36] Other critics noted with irony the ease with which some leading members of the clergy, particularly Peale, utilized the latest in media technology for their evangelistic efforts, pointing to the array of books, television and radio programs, movies, and popular music on the market.[37] It was easy to get the impression that there was something "wrong" with these inhabitants of middle America, that if there were problems in the nation they were not general but localized among the consumers of popular culture.

The learned secular and religious publications pointed to the fact that the revitalization movement had found its appeal in popular culture. Generally they were kinder in their treatment of Graham than of Peale, having a better frame of reference for locating the younger evangelist's theology and approach, polished up and modernized, within an American fundamentalist, revivalist tradition. Pealeism produced for them no similar American associations—except perhaps the secular American dream—and they labored to name the phenomenon. Soon the term that became ubiquitous in referring to Pealeism after 1952 was "cult"; the movement was variously labeled "The Cult of Reassurance," "The Cult of Positive Thinking," "The Cult of Easy

Religion," and "The Cult of Happiness." Although formal definitions of cult usually refer to a personality-based movement, these designations were based on Peale's message.

Perceived in this way, Peale's national following had all the defects of a populist expression, yet some critics managed to concede a few advantages. On the positive side, a few admitted that Peale's success came as a reminder to traditional churches of their shortcomings, particularly as they related to a personal ministry to individuals.[38] They also recognized that the nature of his ministry highlighted the contributions modern psychology could offer to the field of pastoral counseling and the historic cure of souls. But more generally the critics regarded Peale's populist expression as a sad and tragic commentary on the cultural and intellectual bankruptcy of bourgeois life. They did not miss its undercurrent of social leveling or its assertion of entitlement, although those issues were not addressed directly. It was critiqued in other ways. Populist religion, it was said, "democratized God" and ignored the image of a transcendental God as "the Hound of Heaven." It blasphemed God by using popular forms and popular expressions as ways to worship. It forgot the message that people were supposed to be humble, contrite, supplicants of God. The critics tried, largely in vain, to call people back to the religion of the sterner times they themselves remembered.

Repeatedly the critics asked, "What power do these people want?" these supporters of Peale. Responding to their own question, they observed that power had, in fact, become empty, unsatisfying, and actually less available to individuals at that time than earlier in the American past. But if the critics regarded as empty the self-empowerment that the eager purchasers of *The Power of Positive Thinking* desired, the seekers did not share that view. About the many middle-aged women who responded to Pealeism, a critic observed, "They have no projects in which to mix themselves" and "their apathy is their unspoken cry for such opportunities."[39] In an era prior to the organized women's movement, that was undoubtedly true for many women, although it is impossible to know if some may have experienced a greater degree of empowerment through exposure to even some simple Peale strategies, such as coping with the disabling crises produced by a family member's alcoholism, as the woman in the *Guideposts* story claimed.

Although Peale's politics were never mentioned in critiques of the book and his message, there were insinuations of the political divisions that separated the critics from the populist revival in progress. Ecumenism was one such revealing and divisive issue, with theological and political liberals lined up behind the National Council of Churches and conservative evangelicals more supportive of the National Association of Evangelicals. The National Council faced a particularly complicated situation in its relationship with Peale. The Council and Peale were never particularly friendly to each other, both because of the organization's political orientation and its image to Peale of the kind of "big shotism" he deplored. Yet he received Council support for his radio program and quietly accepted Ruth's decision to serve as one of its vice presidents. The tensions that existed on both sides occasionally broke through to the surface, as they did when the chairman of the Council's Division of Christian Education, Paul Calvin Payne, was asked his opinion of *The Power of Positive Thinking*. Payne answered that he felt "negative" about Peale's "easy optimistic sentiments."[40] His published comment left S. Franklin Mack, Director of the Council's Broadcasting and Film Commission, the agency sponsoring Peale's program, to rescue the situation, which he attempted by observing that "it cannot be said of us as it was of Jesus that the common people heard him gladly," although Peale enjoyed rapport "with people in all walks of life."[41]

There was other evidence of the strain that the struggle over ecumenism represented. An additional sign of the tension between the Council and Peale appeared in a report on the first National Convention of Christian Men, another constituent part of the Council. The convention, a meeting of 3,200 men and a gathering of the sort in which Peale flourished, invited the famous author to address the delegates. Included on the program were Graham and Republican Congressman Walter Judd, as well as Eugene Carson Blake of the Council and Quaker Elton Trueblood. When the columnist for *Christian Century* reported on the event he reflected on which speakers might have had the greatest impact on the delegates and how deep was their inspiration. He reluctantly admitted that Blake made "little impression" and Trueblood "failed to receive the attention his merits deserved," while Peale, the writer agreed, fired the crowd with his shouts to "Pray big! Believe big! Act big." They also responded positively, he noted, to Judd's Republican politics and gave Graham a standing ovation.[42] The leadership and the laity of the Council clearly had differing opinions of Peale and his evangelical colleagues.

But the criticisms most wounding to Peale personally were those that scored his theology. He had always felt on the defensive in intellectual debate, apologetic about a sense of limited scholarly abilities, and for that reason may even have deliberately "written down" to telegraph the book's popular style to readers and reviewers alike. The attacks on his popularized psychology were comparatively less hurtful to Peale and were less subject to debate, although many critics described his reliance on the self-hypnotic aspect of positive thinking as potentially dangerous evidence of denial, repression, and automanipulation. The most damaging criticisms of the book referred to it as blasphemous, heretical, and alien to the Christian tradition.

Peale's defense and that of his public supporters were that he was misunderstood: Critics had simply failed to appreciate the orthodox Christianity implicit in the text. In addition, there was the answer provided by results. The message, it was said, "worked," as the testimony of thousands demonstrated. Yet it was difficult for him to argue from within the text for evidence of Christian orthodoxy. Although Jesus was mentioned throughout, it was as

eale's parents, Anna and Clifford Peale. TOP RIGHT: *Peale's grandparents, Samuel and Laura eale.* LOWER LEFT: *The young Norman Vincent Peale.* LOWER RIGHT: *Norman Vincent eale's mother Anna Peale, devout Methodist and devoted mother, managed to steer two of her sons—orman and Leonard—into the Methodist ministry, and the third—Robert—into a career in edicine.*

TOP: *The Reverend Clifford Peale with two of his sons, Robert, left, and Norman, right. His formative years among Ohio Methodists had a lasting effect on Norman.* BELOW LEFT: *Norman (seated) and Robert Peale, ages eight and five. The eldest of the three Peale children, Norman grew up in a close-knit, emotionally intense family. A strong sense of family regard remained with him, finding expression even in the sociological implications of his religious message.*
BELOW RIGHT: *Norman Vincent Peale as a high school graduate, 1916.*

TOP: *Norman Vincent Peale with brother Leonard, father Clifford, and brother Robert Peale.* BOTTOM: *Kappa Sigma Pi was a Brotherhood of St. Paul Boys Club. The photo shows the group meeting at the church of Norman's father. Norman Vincent Peale is third from the left; his father is seated at the right. Activities related to the Methodist church were the centerpiece of the young Peale's social life.*

LEFT: The wedding of Ruth Stafford and Norman Vincent Peale, June 20, 1930, in Syracuse, New York, where Norman was the minister of the University Methodist Church. Ruth had been a student at Syracuse University. Ruth's sister-in-law, Eleanor Stafford, is at the left; Norman's brother Robert is at the right.
BELOW: Family photo taken at farm in Pawling: Ruth Stafford Peale; daughter Elizabeth; son John; and daughter Margaret (c. late 1940s).

Hill Farm; Peale residence, Pawling, New York.

The mature minister as doctor of divinity, about 1950.

The extended Peale family of adult children, their spouses, and grandchildren. Son John Peale is standing at far left; daughters Elizabeth Allen and Margaret Everett are standing, second and third from right.

LEFT: *Ruth Peale outside the New York apartment, on her way to services at Marble Collegiate Church (c. 1970).*

Marble Collegiate Church, New York City.

ABOVE: *A typical Sunday morning scene outside Marble Collegiate Church, as a crowd gathers to gain admittance to the service and hear a celebrated Peale sermon. This photo was probably taken at the height of Peale's popularity in the late 1950s.* BELOW: *Norman Vincent Peale preaches to an Easter Day congregation at Marble Collegiate Church in New York City. From the early 1940s on, Peale attracted capacity audiences at the church. Crowds gathered early on Sunday mornings outside the church to get tickets assuring seats inside the sanctuary. The overflow audience was directed to an adjacent auditorium to watch and participate in the service via closed circuit television.*

ABOVE: *Reverend Norman Vincent Peale preaches at Marble Collegiate Church on the occasion of his 40th anniversary at the church in 1972. Peale always spoke directly to his audience without a lectern or pulpit coming between them. The lectern to his right held the Bible. On the platform with him to the right of the flags is his brother, the Reverend Leonard Peale; his son, the Reverend John Peale; and his associate and eventual successor, the Reverend Arthur Caliandro.*
BELOW: *A business luncheon audience, probably at the Rotary Club of New York City. Peale was always a hit at meetings of sales and business people.*

ABOVE: *Ruth Peale, center, with Lowell Thomas, left, and Dewitt Wallace, right. Thomas, a friend and Pawling neighbor of the Peales, helped to promote a revitalized* Guideposts *magazine. After a fire destroyed the first building housing* Guideposts, *along with its records, newscaster Thomas broadcast an appeal for subscribers to call in their names and addresses. The result was a dramatic increase in subscribers. Dewitt Wallace, publisher of* The Reader's Digest, *was another long-time Peale friend and supporter of* Guideposts. BELOW: *An award dinner for the American Foundation for Religion and Psychiatry (now known as the Institute for Religion and Health) on April 27, 1970. The Foundation had its origins in the 1930s, a shoestring operation conducted in offices at Marble Collegiate Church by Peale and his Freudian-trained colleague, Dr. Smiley Blanton. The I.R.H. has become nationally and internationally renowned as a network of training centers for therapists and clinicians with a background in pastoral care.*

ABOVE: *Architect's rendering of expansion of the Foundation for Christian Living, Pawling, New York, c. 1977.* BELOW: *Entrance to the Peale Center for Positive Thinking, the latest addition to the Foundation for Christian Living (renamed the Peale Center in 1991) in Pawling, New York.*

Norman Vincent Peale with President Dwight D. Eisenhower. Peale had campaigned for the Eisenhower-Nixon ticket and remained close to both families, especially the Nixons.

Peale on the stump with a secular audience, 1949.

lthough essentially a life-long Republican, Peale liked to identify himself as a political independent. e appears here with a group of clergy, meeting with President Truman.

ABOVE: *Norman and Ruth Peale with President and Madame Chiang Kai-shek at the President's summer place. Like many others during the Cold War, Peale supported Chiang Kai-shek and opposed communist mainland China.*
BELOW: *Peale with President and Mrs. Richard M. Nixon, and David and Julie Eisenhower. The Reverend Peale conducted the wedding for David Eisenhower and Julie Nixon in December 1968. Nixon attended services at Marble Collegiate Church on numerous occasions.*

RIGHT: *Peale visited Vietnam at the request of President Nixon (c. 1969). He quietly supported Nixon's foreign policy.* BELOW: *Norman and Ruth Peale presenting the NVP Award to President Ronald Reagan, September 6, 1990.*

ABOVE: *The Peales meeting with Pope John Paul II in Rome, early April 1984.* BELOW: *The Great Communicator, Norman Vincent Peale, in action with his ministries.*

friend and as Way-Shower in the New Thought sense and not as the incarnational Christ. The approach of the work did not admit a place for a discussion of flawed human nature, nor the need for sacrifice and service.

The most serious allegation raised by critics, then, was not that Pealeism was a secular movement masquerading as a religious one but rather that it was a kind of shadow religion, a distorted and dangerous adaptation of the real thing. Peale's cosmic and transcendental concept of divinity was viewed as blasphemous because it seemed to reduce God to human proportions and to make God the servant of humanity. The heart of the argument was Peale's basic identification of his work as practical Christianity. Was the gospel he preached practical and functional only? Was it a means to an end, or was it an end in itself? Did positive thinking aim to make people better, or did it promise to make them the best? Was the practical side of his message ancillary to the basic Christianity, or was it the central concept?

To the critics, the answers were obvious. For some of the harshest critics, Pealeism represented the worst aspects of the populist revival, his best-selling book the evidence for an almost demagogic appeal to empower the uninformed with dangerous notions. Softened by the siren songs of popular culture, middle-brow Americans were conspiring with their own self-delusion.

On balance, it is probably true that the book addressed a recently awakened desire among middle Americans for a better quality of life in all areas: health, relationships, finance, psyche. To that end, it focused on the individual and nourished a form of narcissism that could sometimes be beneficial. Yet it challenged prevailing conventional definitions of religion to regard as religious a movement that endorsed instrumental values as an end in themselves. It accepted the culture uncritically, emphasized religious autonomy over religious community, and focused on religious awareness rather than theological understanding.

More positively, Peale's book signaled at least temporarily an end to religious apathy. And as even its critics agreed, it persuaded mainstream churches to re-examine their ministry to individuals and to consider more modern approaches to the cure of souls.

For Peale, his personal crisis over the book's reception occurred in September 1955, just as he and Ruth returned from another summer in Europe. William Peters' essay, "The Case Against 'Easy' Religion," with sharp attacks on the work by leading churchmen, had just appeared in *Redbook*. Family and associates thought they might be able to spare Peale its indictments until he had been prepared for its contents. But shortly after his return, he was called to visit his terminally ill father in Harrison Valley, Pennsylvania, and alone on the train, he picked up Peters' article. The essay was filled with the familiar stinging rebukes, about using religion and constructing a cult and circulating heretical ideas, but on this occasion the words cut particularly hard because the authors quoted were well-respected leaders in the church. Peale decided he had had enough and drafted a letter of resignation to Marble Collegiate Church. He planned to leave Marble Church, but not the ministry, persuaded that "I can do better outside of the organized church. I'll set up something of my own."[43]

In fact, he remained at Marble Church for almost another thirty years. Urged by Ruth to hold off on submitting the resignation letter, he soon heard from his stepmother that his father's final request had been that his eldest son remain in the church, that "Peales were not quitters." It was in their last conversation that his father had reminded him of the unique, syncretic quality of his ministry, formed, he said, out of "positive thinking, Methodist witnessing, Baptist preaching, and Unity." It was presumably this parting expression of concern that convinced the son to remain and carry on the work.[44] Surely other factors entered into his reconsideration, but they can only remain speculations. He stayed at Marble while channeling his talents and resources into the Foundation for Christian Living.

Revival and Redirection

Sydney Ahlstrom once observed that Peale "was as important for the religious revival of the fifties as George Whitefield had been for the Great Awakening of the eighteenth century."[45] While Peale would have preferred a comparison to John Wesley, the qualities of religious itinerancy, of energizing crowds on the social periphery, of simplifying the way to faith, of helping reorder old lines of religious affiliation were as evident in Peale as in Whitefield. Pealeism alone, however, did not fuel the revival, which was carried forward as much by Billy Graham as by Peale. Just as Peale learned to adapt and restate the tenets of religious liberalism, Graham found new ways to advance an updated fundamentalism.[46] Never particularly close, the two men had their most substantial contacts in the years between 1953 and 1960. By 1960, the revival energy spent and the evangelical political thrust temporarily blunted by its embarrassing intrusion into the Kennedy campaign, the two men went their separate ways, to meet again when celebratory political events placed them on the same platform.

While Peale was regularly increasing his national audience through his prodigious cross-country speaking efforts, Graham was giving new meaning to evangelistic meetings. Working through the Billy Graham Association, the Graham model of the Foundation for Christian Living, the handsome southern preacher was drawing thousands to his urban revivals. He and Peale seemed to have little in common as the revitalization movement got under way in 1950, one a groomed neofundamentalist and the other a cheery neomodernist. Yet in the years when the movement was gathering steam, they developed a courteous relationship, initially born out of their common contacts within the National Association of Evangelicals and their shared sympathy for conservative politics.[47]

Previously Peale had not been known to cultivate acquaintances with fundamentalists. His interactions with them had been almost as difficult as those with liberals, with the conservatives proving more zealous and persistent in writing him personal letters of sharp rebuke for his lack of religious orthodoxy.[48] But it was a measure of the fundamentalists' relative social power, as well as Peale's own sense of status security in relation to them, that allowed him to feel less threatened by them and even at times to respond to their needs. Almost every summer, for example, he accepted the invitation to preach to the many religious conservatives—evangelicals and fundamentalists—gathered in Ocean Grove, New Jersey, for an inspirational camp meeting experience. Privately, however, Peale was intellectually far more alienated from fundamentalism than he was from liberalism. Years later when he recalled his association with the youthful evangelist twenty years his junior, a sense of distance remained: We were "friends, not bosom companions," he said, who "worked the same street, but from different sides."[49] He had been a force on the national scene for at least ten years when Graham began his crusading efforts in 1949.

As a Graham biographer observed, theological distinctiveness has not been a characteristic of American revivalism, which depends for its success on the ability to appeal to the widest possible audience.[50] When Graham first came to the attention of Peale, as of others, he was an ordained southern Baptist preacher, theologically premillennialist and politically conservative. Even in 1956 he continued, if defensively, to say of his views, "if by fundamentalist you mean a person who accepts the authority of the Scriptures, the virgin birth of Christ, the atoning death of Christ, his bodily resurrection, His second coming, and personal salvation by faith through grace, then I am a fundamentalist."[51] It was not a perspective publicly shared by the author of *The Power of Positive Thinking*, whose comment when he first heard of the young preacher was that he seemed to be "of the sensational type which I never took to much myself."[52] Still, he admitted to the *Guideposts* editor who was considering a story on Graham, he thought he was "accomplishing things, and I think we ought to watch him with an open mind."[53] Then when Graham's own 1953 book *Peace with God* appeared, which revealed not only soft edges to his hard-core orthodoxy but also areas of overlap with such Peale themes as "peace of mind" and "serenity," there was room for claiming fresh opportunities for theological dialogue.

The relationship between the two men first appeared as a friendship in 1955, as Graham began planning for his New York City crusade. His most elaborate project to date, the campaign depended on the support of local ministers, a requirement imposed by Graham despite conservative claims that such a policy forced him to compromise his theology for the likes of Peale. Graham's rallies, which made him a media attraction the equal of Peale, were producing thousands of "decisions" for Christ. As with Peale's audiences, the majority of people making "decisions," between 65 and 75 percent, were already church members.[54] More than half the converts, or over 60 percent, were women, but unlike Peale's supporters, they were young: 45 percent were under eighteen, and another 20 percent were between nineteen and twenty-nine. It was obvious that neither Graham's audience nor Peale's consisted of many unchurched or "come-outers." Like Peale's constituency, they were church members and either seekers after alternatives or simply curious.

The first meeting of substance between Graham and Peale took place during the summer of 1955 in Switzerland. An inveterate traveler, Peale had earlier kindled a fondness for the physical beauty of Switzerland as well as its other rewards, and Graham had come to enjoy the country as well. The two men met and talked for three hours about Graham's plans for his New York crusade.[55] As a major figure in religious life, in the nation as well as New York City, Peale could prove vital to the crusade's success. An earlier Graham plan for a series of mass meetings in 1953 had to be scrapped for lack of support from area churches.[56]

That conversation presumably changed Peale from a Graham skeptic to a supporter. When he learned the following year that Graham had received the Clergy Churchman of the Year Award, Peale, never given to understatement, and frequently the most enthusiastic cheerleader on the court, commented to his younger acquaintance that he was "the greatest living preacher of the Gospel of Jesus Christ."[57] During the long months of preparation preceding the crusade, the two had numerous opportunities to get better acquainted as Peale served on a blue ribbon planning committee with other local clergy, prominent civic leaders, and a few notable liberals, such as Henry Pitney Van Dusen of Union Theological Seminary, who had himself been converted during a Billy Sunday crusade.

Broadcast on television every Saturday night, Graham's New York meetings, which ran for ninety-seven nights, secured his position alongside Peale's as a national religious luminary. Peale was a vocal and energetic supporter of the crusade. From his point of view, and surely from Graham's, the campaign was a ringing success: A total attendance of over two million people produced 56,767 "decisions for Christ."[58] Although a *Life* magazine spotcheck revealed that four-fifths of those decisions were made by people who were already church members, and although not a few ministers worried that lasting effects would prove negligible, most members of the clergy were satisfied that Graham's efforts would bring new life to the churches.[59] Peale was particularly pleased with the results of the meetings, especially since his church received the largest number of converts.[60]

An event with lasting cultural significance occurred the night Peale joined the rest of the mourners to make another "commitment" to Christ. Received by Graham, Peale described it as a "second blessing" when he "went deeper" into the faith.[61] A later commentator called it "The first major interface between revivalism and New Thought," acknowledging that for Graham it represented the "recognition of Peale as a fellow born-again Christian . . . despite his Pelagianism" with its sense of self-empowerment.[62] It was for

Peale an outward manifestation of what had long been part of his personal history, namely, the connection between harmonial mental science beliefs and traditional evangelicalism. Symbolically, in case supporters had any doubt, it appeared to bring the left and right theological wings of the revitalization effort together. Theological liberals, however, would have had great difficulty imagining Peale occupying any ideological left, political or theological.

In the five years after the New York meetings, their images ever before the public, Peale and Graham discovered mutually supportive friends. His relationship with Graham may have brought Peale closer within the charmed circle of the National Association of Evangelicals (NAE). He had already developed a productive friendship with one of the Association's chief benefactors, J. Howard Pew, through their partisan activities with Spiritual Mobilization and the Christian Freedom Foundation. It was Graham, however, who had the longer history with Association stalwarts Carl Henry, Harold J. Ockenga, and L. Nelson Bell. Political similarities rather than theological affinity finally allowed the NAE to regard Peale as one of them. Like Graham, Peale shared the NAE's conservative agenda on such issues as labor, anti-Communism, Red China, and containment. It was those Cold War issues that had first brought Peale and Pew together, with Pew serving as one of the important early contributors to *Guideposts*. It was also quietly reported that Pew was the financial backer of the NAE's publication, *Christianity Today*, which first appeared in 1956.[63] An orthodox Presbyterian, Pew presumably would have little in common with Peale's theology as it was presented in *The Power of Positive Thinking*.

There was also the ability that Peale and Graham shared to mobilize the latest media technology, to press it into the service of the revival's evangelistic effort, which was as vital to the movement's growth during the decade as their political perspective. If freighted with theological differences, the revival was lifted on the wings of modern media developments and the fellowship of political partisans on the populist front. Coordinating Graham's resources was the Billy Graham Evangelistic Association in Minneapolis, Minnesota, then a technologically more advanced operation than Peale's Foundation for Christian Living, and the central agency for managing all his enterprises. It was the planning hub for his campaigns, and it handled all his direct mailings and request letters. As Peale's resources increased, the Foundation for Christian Living would come close to approximating Graham's operation. Together the two independent agencies were in contact with a vast cross-section of evangelical Protestants.

By 1960, with revitalization efforts exhausted and evangelical politics temporarily discredited, Graham and Peale turned to their separate national ministries. As revival fires cooled, within the institutional church poll takers and pulse readers searched for reasons to explain the transitory explosion of faith. There were those who concluded that the image of growth had been just a mirage; others claimed attrition within Roman Catholicism accounted for the smaller numbers; still others argued that the time had arrived for the periodic winnowing of dead wood, which would leave the faithful remnant intact. What the evidence revealed was that between 1952 and 1971 church growth was largely confined to conservative denominations, with some groups showing dramatic percentage increases. For example, during those years the Church of God (Cleveland), a Pentecostal group, increased 120.9 percent; the Church of Jesus Christ of Latter-day Saints, or Mormons, by 98 percent; and Graham's Southern Baptist Convention by 45.5 percent. By comparison, the Episcopal Church grew by only 19.2 percent; the Presbyterian Church, U.S.A., by 8.8 percent; and the Methodist Church by only 5.2 percent. Even within those bodies it was the more conservative faction that was largely responsible for the increase.[64] Peale's audience continued to grow after 1960, but not as dramatically as in the previous decade, when the publication of *The Power of Positive Thinking* coincided with revival energies.

The revival had revealed a number of important cultural realities. It showed the resurgence of a trend toward conservative and evangelical faith. But the disaffection from mainstream churches did not inevitably turn toward Graham and traditional religious conservatism. Many continued as Peale supporters while remaining within their local churches: If their loyalties were not in conflict, their resources were surely divided as they contributed to two religious organizations where previously there had been one. One important conclusion to be drawn from the success of Peale's message was that there continued a strong tradition within popular culture of a belief in religion validated by nature, or even Emerson's Nature. While members of the intellectual community tried to reassure themselves that reliance on a basic, or primal, belief in Nature as divine revelation had been banished from popular culture at least since the Enlightenment, its repeated resurgence, as in the fifties revival, gave the lie to their hopes. For many middle Americans, the abstractions of systematic theology carried little weight, especially in times of crisis.

The marginal, unchurched members of Peale's national audience, attracted by a belief structure that was neither demanding nor doctrinaire, were likely recruits for America's conservative civil religion. But the revitalization movement had not been just a negative vote on mainstream religion, with its liberal theology and left wing politics. It had also been a positive expression of a group's desire for recognition, political and economic entitlement, and a stronger voice in their own religious destinies. Additionally, it was a generalized statement about the uncertainty and anomie of the contemporary culture and a reaching out for a tie that would bind old roots with new growth.

The Foundation for Christian Living

The Foundation for Christian Living (FCL; since 1991 the Peale Center for Christian Living) was transformed into the nerve center of Peale's independent ministry during the decade of the fifties. In 1950, at the start of the "Peale decade," it had a mailing list of 50,000; ten years later, due to Peale's popular image and the sales of *The Power of Positive Thinking*, its subscriber list had grown to 325,000. After that, growth remained steady and impressive: a mailing list of 500,000 in 1970, 700,000 in 1980, and 800,000 in 1989.[65] Peale believed that the FCL, like *Guideposts*, had "found a need and was filling it." But it was also the case that Peale had focused his energies on the growth of the Foundation, endorsing new programs and channeling the resources of his speaking tours and benevolent contributions to their success.

The real power at FCL, however, was Ruth Peale. The organization had grown out of a sermon outreach program at Marble Collegiate Church that she had created in the early 1940s, along with the assistance of volunteers who helped her circulate her husband's sermons. Metamorphosed over time into a far more expansive operation, it eventually became a state-of-the-art facility in Pawling, New York. Some fifty years later, in 1989, it was home to the new Peale Center for Positive Thinking, with a staff of over a hundred, the latest in electronic technology, a diversified service program, a mailing list of a million—Peale's actual constituency at the time—and a budget of $14 million. Peale consistently referred to his work as a "team ministry," and nowhere was that more evident than at FCL, where Ruth, the CEO, managed most of the executive operations singlehandedly.

The Sermon Publications Project was moved from Marble Church to Pawling in 1948, opening up new possibilities for greater independence for Peale's public ministry. With the new arrangement, Sermon Publications and *Guideposts* shared facilities in the small upstate village. The location allowed Ruth more flexibility in dealing with the dual demands of a busy household and her increasingly complex commitments to her husband's wider ministry. As often as possible, the Peale family of five retreated to their Pawling country home on Fridays and Saturdays when Norman was in town, and Ruth used some of that time to tend the affairs of Sermon Publications, renamed the Foundation for Christian Living in 1956. By that time, when the Foundation moved into its own quarters and separated its operations from those of *Guideposts*, Peale had firmly established a public ministry independent of his role at Marble Collegiate Church. Some officers of Marble Church served on the FCL board of trustees, and there was an overlap with a few of the staff. But the FCL parish was distinct from Marble Church, its members resembling to a large extent the characteristics of the *Guideposts* subscription list.[66]

Pawling was an especially congenial place for the Peales. It offered privacy, a small-town atmosphere, and a quality of life more congenial to them than Manhattan, where they often seemed a kind of misplaced Grant Wood couple. Peale referred regularly to his "farm" in the country. In fact, Pawling, and particularly Quaker Hill where the Peales had their home, was the residence of some of the nation's wealthiest citizens, including, at the time, former New York Governor Thomas E. Dewey as well as radio commentator

Lowell Thomas. Peale studiously avoided any semblance of pretentiousness when referring to his place, making it sound little more than a cottage stuck on someone's back forty. Rich themselves by any standards by the midfifties due to the royalties from the books, the Peales continued to reflect the personal habits and fiscal mentality of parsonage-trained preachers' kids. They found more appeal in their country hideaway's charm than in New York's cosmopolitan smorgasbord of theaters, museums, concerts, and films. Although they wore fine clothes and drove luxury cars, the most obvious evidence of their financial privilege was their extensive foreign travel.

Harvesting the work begun by other Peale ventures, the Foundation grew rapidly in its early days.[67] It made no direct appeals for funds, readily sending its literature freely, securing its income from other branches of Peale's public ministry and through his business and political contacts. Available evidence showed that, generally, a couple of dozen large contributors produced the resources for half the Foundation's operating budget, with the remainder supplied by the minimal contributions of a large scattered following and Peale's personal fund-raising efforts.[68] That a small group of financial angels supported a needy client base was hardly unusual among private philanthropic organizations.

As the headquarters for Peale's independent ministry, the Foundation managed his extensive public speaking commitments, especially the secular ones to business and civic groups. The self-declared "missionary to American business," whom critics accused of salving the conscience of the nation's business community, Peale had cultivated through these contacts important benefactors for his various projects—men like Pew and Stanley S. Kresge, J. C. Penney, Walter Teagle, William Danforth, Judson Sayre, E. F. Hutton, Dewitt Wallace, Walter Annenberg, and J. Clement Stone. Speeches before service groups and trade organizations were another important source of revenue.

Free to mold this independent ministry to his own design, Peale created the FCL as the visible manifestation of his positive thinking message. The FCL's slogan, printed on all its literature, was "The Advancement of Christianity as a Practical Way of Life." All of its programs, from its Positive Thinkers Club to its counseling service to distraught letter writers to its own inspirational magazine *Plus,* were developed around the concept of practical Christianity synonymous with Peale's ministry. Theologian Paul Tillich once tried to point out to Peale what he saw as the difficulty inherent in a ministry to the business community, which, said Tillich, used "religion to readapt people to the demands of industrial society." It was wrong, Tillich thought, to invoke a belief that "teaches executives of corporations, for example, who cannot continue in their jobs, how to adjust."[69] Peale answered Tillich by denying rather than refuting his claims. When Tillich responded to Peale again he noted that earlier statements by him about Peale had been misquoted, but that he continued to believe that there was "a very great danger" in making a utilitarian purpose "the primary function of religion."

Peale could offer a quite different and much fuller response to a question about the nature of his ministry when the source was a friendly writer from *Guideposts* rather than a suspected critic from the academy. In answer to a question from John Sherrill of *Guideposts* about where he stood on the theological spectrum, Peale used language reminiscent of Graham: "I'm a conservative, and I will tell you exactly what I mean by that. I mean that I have accepted the Lord Jesus Christ as my personal Savior. I mean that I believe my sins are forgiven by the atoning work of grace on the cross." Then he added, "Now I'll tell you something else. . . . I personally love and understand this way of stating the Christian gospel. But I am absolutely and thoroughly convinced that it is my mission never to use this language in trying to communicate with the audience that has been given me."[70] Presumably, Peale felt the traditional and conservative language could be divisive and alienating to those who responded to his personal ministry.

The FCL continued to grow. Even as he aged, Peale continued his extensive speaking commitments, producing revenues for the Foundation, adding new programs and dropping unproductive ones. The volume of mail, most of it still written by women, kept increasing. New technology and additions to the physical plant in Pawling kept pace with continued growth. Gifts from donors facilitated expansion of the plant in 1956 and again in 1969 and 1979, and then a major addition in 1987 of the Peale Center for Positive Thinking, an impressive gray stone structure complete with a library and the technology for making television videos.

The chief ministry of the FCL over the years had been the production and distribution of inspirational literature along evangelical New Thought lines. That service was updated with the last addition to the plant, when production facilities were added which allowed for the development of films to be marketed as videocassettes. The Foundation's own inspirational pamphlet, *Plus*, a collection of three Peale sermons, was also given a revamped look when the FCL acquired a new editor, Eric Fellman. Fellman replaced the first editor at the FCL, Myron Boardman, who had been head of general interest books at Prentice-Hall before joining the Foundation, where he coordinated all the publishing activities until his retirement in 1984. Fellman was a young evangelical graduate of Wheaton College, previously the director of the *Moody Monthly*, and the person with major responsibility for the publication of *Plus*. The change was indicative of Peale's own priorities at the time.

But the most innovative feature added to the Foundation's program was the production of original films. The effort was managed by Positive Communications, Incorporated (PCI), a wholly owned subsidiary of the Foundation, endowed largely by the Peales themselves. PCI created and planned to distribute motivational videotapes, based on Peale's positive thinking message, targeted for an elementary school population in kindergarten through grade five, an age group assumed to be at highest risk for challenges to self-esteem. Unlike earlier motivational films, however, Peale did not personally appear in the PCI productions and was referred to only in passing.[71] Modeled after such well-known children's programs as "Sesame Street" and "The Electric Company," the tapes were artfully crafted under the direction of a former Lorimar Productions executive, who had himself been persuaded of the value of Peale's message.[72] The videos were intended to be coordinated with a curriculum guide that would involve the participation of classroom teachers and parents. In its initial offering, the PCI program—tapes, study guides, and such—was adopted by the entire school system of West Virginia. One rationale for the tapes, said the producer, was "to convert Dr. Peale's personality into a program."[73]

That had actually been the thrust of Peale's entire public ministry since his first arrival at Marble Church. His message was a reflection of his many-sided personality: his own expansive optimism and experience with good fortune, his combativeness, his dark fears and insecurity, his steady reiteration that he preached out of his own needs. It was this appearance of an accessible Everyman, manifest in his message, that drew him close to his public, sold his books, and cast him in the forefront of the fifties revival. And it was the phenomenal success of *The Power of Positive Thinking* that determined his role in the revival.

Notes

1. Edmund Fuller, "Pitchmen in the Pulpit," *The Saturday Review*, March 9, 1957.

2. Emil Coué, a French pharmacist and writer of inspirational tracts who toured the United States in the twenties, was best remembered for his saying, "Every day in every way I'm getting better and better."

3. New York *Sunday News*, December 11, 1955; and William McLoughlin, *Revivals, Awakenings, and Reform* (New York: Ronald Press, 1959).

4. Ibid.

5. See Paul Hutchinson, "Have We a 'New' Religion?" *Life*, April 11, 1955.

6. William Lee Miller, "Some Negative Thinking about Norman Vincent Peale," *The Reporter*, January 13, 1955.

7. Ibid.

8. See Arthur Gordon, *One Man's Way: The Story and Message of Norman Vincent Peale, Minister to Millions* (Pawling, NY: Foundation for Christian Living, 1972), Chap. XV.

9. These numbers were taken from a graph in A. Roy Eckardt's *The Surge of Piety in America* (New York: Association Press, 1958), p. 22.

10. See McLoughlin, *Revivals, Awakenings, and Reform*. The anthropological concept of a religious revitalization movement was developed by Anthony Wallace in his treatment of Handsome Lake and the Seneca Indians.

11. The three organizations within Peale's audience that gathered evidence on his supporters were Marble Collegiate Church, *Guideposts*, and the Foundation for

Christian Living. In their separate profiles of the Peale audience there is a marvelous redundancy: Over half his supporters (and three quarters at *Guideposts* and FCL) were women, generally middle aged, and with an evangelical Protestant background. There is no demographic profile of his radio or television audience or of the purchasers of his books. Surely there was a significant number of "unchurched" to be found there.

12. R. Laurence Moore, *Religious Outsiders and the Making of Americans* (New York: Oxford University Press, 1986), p. xii.

13. "Minister to Millions" is the subtitle of Arthur Gordon's biography of Peale.

14. Hutchinson, "Have We a 'New' Religion?" *Life*, April 11, 1955, p. 148.

15. See Edward A. Tiryakian, "Toward the Sociology of Esoteric Culture," in Tiryakian (ed.), *On the Margin of the Visible: Sociology, the Esoteric, and the Occult* (New York: John Wiley & Sons, 1974), p. 273.

16. For ideas about interest-group affiliations I am indebted to Robert Wuthnow, *The Restructuring of American Religion: Society and Faith Since World War II* (Princeton, NJ: Princeton University Press, 1988).

17. In "Recent Trends in Church Membership and Participation: An Introduction," the authors note: "Juxtaposing these trends [outward growth and doctrinal decline] with those previously outlined for church membership and participation presents a number of interesting contrasts. Most obvious is that the two sets of trends moved in opposite directions during the 1950s and early 1960s. Traditional Christian beliefs and the saliency of religion eroded, while church membership and participation grew." David A. Roozen and Jackson W. Carroll, in Dean R. Hoge and David A. Roozen (eds.), *Understanding Church Growth and Decline: 1950–1978* (New York: Pilgrim Press, 1979), p. 38.

18. See James H. Smylie, "Church Growth and Decline in Historical Perspective: Protestant Quest for Identity, Leadership, and Meaning," in Hoge and Roozen, *Understanding Church Growth and Decline*, p. 70.

19. See Wuthnow, *The Restructuring of American Religion*, Chap. 3.

20. Sydney Ahlstrom, *A Religious History of the American People* (New Haven, CT: Yale University Press, 1972), p. 962.

21. McLoughlin, *Revivals, Awakenings, and Reform.*

22. See Wuthnow, *The Restructuring of American Religion*, Chaps. 2, 3.

23. Herberg argued that statistically and symbolically most Americans in the 1950s were third-generation immigrants, who found in religion the identity that eluded them elsewhere in the culture. See Will Herberg, *Protestant, Catholic, Jew* (New York: Doubleday, 1955).

24. This excerpt from a review by Sterling North, "Reviewing the Books," appears in the authorized biography by Arthur Gordon, *One Man's Way: The Story and Message of Norman Vincent Peale, Minister to Millions* (Pawling, NY: Foundation for Christian Living, 1972), p. 230.

25. J. Gordon Melton, *The Encyclopedia of American Religions* (Wilmington, NC: McGrath, 1978), Vol. 2, p. 56. Melton locates New Thought within the category of the "Metaphysical Family." Others who noted the New Thought connection in Peale's work include Donald Meyer, *The Positive Thinkers* (New York: Pantheon Books, 1980), an intelligent but sharply critical account; Charles Braden, *Spirits in Rebellion* (Dallas: Southern Methodist University Press, 1963), which points out that

The Power of Positive Thinking "is a distinctly New Thought title," and Richard M. Huber, *The American Idea of Success* (New York: McGraw-Hill, 1971), which argues that "Peale made no contribution to the basic design of New Thought theology, except to dress it up in a particularly attractive package" (p. 321).

26. Interview with NVP, July 6, 1989, Pawling, New York. He made the comment in reference to an article that appeared in an Ernest Holmes publication.

27. Norman Vincent Peale, *The Power of Positive Thinking* (New York: Prentice-Hall, 1952), p. 174.

28. NVP to Samuel Williamson, Dec. 12, 1945; NVP Ms. Coll.

29. Peale, *The Power of Positive Thinking*, p. viii.

30. J. Stillson Judah, a student of metaphysical beliefs, concluded that groups which belong to that circle of belief share fifteen common assumptions. Judah described these as revolt against creedal authority; belief in the divinity of the "inner man"; a concept of God as Divine Mind or Universal Mind; a quasi-gnostic view of God that slides into monism; Jesus as the Christ Principle or "way-shower"; an omission of reference to sin; belief that evil and sickness are unreal and an error of mind; an attachment to pragmatism; an emphasis on self-realization; religious experience as scientifically verifiable; an acceptance of a psychological approach; belief in optimism; an acceptance of prosperity; a belief in the "inner meaning" of words; and belief in healing through the mind. Judah, *The History and Philosophy of Metaphysical Movements in America* (Philadelphia, PA: Westminister Press, 1967), p. 305. Although most of these characteristics apply to *The Power of Positive Thinking*, ten of the themes recur throughout the book.

31. Liston Pope wrote to Peale's friend and biographer Arthur Gordon to correct what Pope said were Gordon's inaccurate characterizations in an article and in the biography of Pope's statements about Peale. Gordon had said that Pope spoke "with searing contempt of what he called the 'peace of mind cults.'" In Pope's objection, he said that while it was well known that he did not agree with "the new religions of reassurance," a careful reading of his speech will make clear that he did not speak of Peale with "searing contempt." Liston Pope to Arthur Gordon, Jan. 15, 1959; NVP Ms. Coll.

32. Both Ruth and Norman Peale said they suffered from a personal crisis generated by the *Redbook* article: William Peters, "The Case Against 'Easy' Religion," *Redbook*, September 1955. They were also unsettled by the article by William Lee Miller, "Some Negative Thinking about Norman Vincent Peale," *The Reporter*, January 13, 1955. In fact, because the onslaught was so great, it was probably the cumulative effect of the negative criticism that finally mattered, although they claimed to be less troubled by the sophisticated and select journals of opinion.

33. Interview with NVP and RSP, February 24, 1981, Pawling, New York. Long familiar with the emotionalism that can be generated in Methodist evangelistic meetings, Norman Peale could come easily to tears. Ruth Peale, by contrast, usually appeared more controlled.

34. Miller, "Some Negative Thinking about Norman Vincent Peale"; Donald Meyer, "The Confidence Man," *New Republic*, July 11, 1955; Curtis Cate, "God and Success," *Atlantic Monthly*, April 1957.

35. Cate, "God and Success"; Meyer, "The Confidence Man."

36. Meyer, "The Confidence Man."

37. Chester Morrison, "Religion Reaches Out," *Look*, December 14, 1954.

38. Hutchinson, "Have We a 'New' Religion?" Hutchinson's long article in *Life* was the most balanced of the Peale critiques.

39. Meyer, "The Confidence Man."

40. *Newsweek*,, February 21, 1953.

41. Ibid.

42. *Christian Century*, October 3, 1956.

43. Interview with NVP, February 23, 1981, Pawling, New York.

44. Ibid. See also Braden, *Spirits in Rebellion*, which alters the ingredients a bit. It is a significant story for Peale, one that he has recounted often and that has appeared in a number of places, often with some variations in the constituent parts.

45. Ahlstrom, *Religious History*, p. 956.

46. Useful discussions of Graham's efforts to adapt his fundamentalist background to the "new evangelicalism" in the 1950s are contained in George Marsden, *Reforming Fundamentalism: Fuller Seminary and the New Evangelicalism* (Grand Rapids, MI: Eerdmans Publishing Company, 1987), pp. 167ff.; and William Martin, *A Prophet with Honor: The Billy Graham Story* (New York: Morrow, 1991), esp. pp. 226–30.

47. See McLoughlin, *Billy Graham, Revivalist in a Secular Age* (New York: Ronald Press, 1960). Not a friendly biography, McLoughlin's work is useful for its inside look at the Graham organization.

48. There are many letters of brutal criticism from fundamentalists and very orthodox evangelicals in the Norman Vincent Peale Manuscript Collection.

49. Interview with NVP, June 17, 1987, Pawling, New York.

50. McLoughlin, *Billy Graham*, pp. 70, 71. See also the studies previously cited of Graham's ministry by Marsden and Martin.

51. *Look* magazine, February 7, 1956; quoted in McLoughlin, *Billy Graham*, p. 70.

52. Privately, however, Peale was willing on occasion to describe himself in language similar to this invoked by Graham; see page 151 for comparison.

53. NVP to Grace Oursler, Apr. 1, 1950; NVP Ms. Coll., Box 756.

54. McLoughlin, *Billy Graham*, pp. 182, 183. McLoughlin also says that the statistics suggest that 55 percent of the "converts" were not only church members but people who had had a "born again" experience.

55. NVP to the Rev. Dan Potter of the Protestant Council of the City of New York, Nov. 16, 1955; NVP Ms. Coll. In his letter Peale spoke of the "three hour visit . . . last summer" with Graham in Switzerland.

56. McLoughlin, *Billy Graham*, p. 148; also Marshall Frady, *Billy Graham: A Parable of American Righteousness* (Boston, MA: Little, Brown, 1979), pp. 292–305. Frady, like McLoughlin, is critical of Graham. More recent accounts, such as Marsden's discussion of Graham's role in the new evangelicalism and Martin's biography, are much more favorable in their judgment of Graham.

57. NVP to Billy Graham, Mar. 24, 1956; NVP Ms. Coll., Box 352.

58. McLoughlin, *Billy Graham*,, p. 149.

59. Ibid., p. 186.

60. Martin, *Prophet with Honor*, p. 238. See also Richard Quebedeaux, *By What Authority: The Rise of Personality Cults in American Christianity* (San Francisco, CA: Harper & Row, 1982), p. 44.

61. NVP to James E. Carter, Mar. 18, 1960; NVP Ms. Coll., Box 333. In his let-

ter Peale wrote: "I had the new birth experience before Billy Graham was born, but in his meetings here I received a second blessing. I believe that conversion may be a progressive experience. Under Billy, to whom I am devoted, I progressed, or went deeper."

62. Quebedeaux, *By What Authority*, p. 44.

63. On Pew and *Christianity Today*, see Ernest Pickering, *Should Fundamentalists Support the Billy Graham Crusades?* (Chicago: 1958), p. 17; and William E. Ashbrook, *Evangelicalism: The New Neutralism* (Columbus, OH: 1958), pp. 23, 24.

A measure of the guardedness involved in developing these partisan activities was revealed in a carefully worded letter from Spiritual Mobilization's director, James Fifield, to Pew. The California clergyman said that in his judgment it was crucial for Pew to gain many other supporters besides himself for his "program." Past experience had demonstrated, said Fifield, that "Unscrupulous leftist elements could make capital out of the fact that you are so largely supporting the enterprise." Because of the cautious nature of the letter, it is difficult to figure out to which of Pew's many projects it referred. James Fifield to J. Howard Pew, Aug. 10, 1950; NVP Ms. Coll.

64. *Yearbook of the American Churches*, 1965, p. 280; 1970, p. 227; 1975, pp. 262–65; Dean M. Kelley, *Why Conservative Churches Are Growing: A Study in the Sociology of Religion* (New York: Harper & Row, 1972), pp. 20–31.

65. According to the secretary to Eric Fellman, publisher of FCL's *Plus* magazine, the magazine in October 1991 had a mailing list of over two million in 118 countries. It is sent free to anyone who requests it. Telephone conversation, October 15, 1991. The mailing list for FCL, however, is listed as a million. See "Peale Center for Christian Living," *Plus*, Special edition, 1991.

66. The NVP Manuscript Collection contains packed files on the work of FCL, and those files in the collection grow larger annually as new FCL material is added. The vast majority of the material consists of "request" and "success" letters.

At FCL, the director of guidance reported that the new profile of letter writers appeared in the 1950s, when middle-aged women became the predominant category. The director said that next to health needs, the women, many of whom were widows, spoke of their loneliness. Interview with Carol Porter, Director of Guidance, FCL, February 23, 1981, Pawling, New York.

67. NVP to Joe Rukenbrod, June 13, 1950; NVP Ms. Coll. Also "A History of the Foundation for Christian Living," typescript of copy placed in the cornerstone of the new Peale Center for Positive Living, June 26, 1987.

The FCL mailing list grew from 50,000 in 1950 to 200,000 by 1953, another testimony to the sales of *The Power of Positive Thinking*.

68. Report of "Sermon Publication Committee" (filed under R. Sheppard), n.d., probably 1946; NVP Ms. Coll. Also conversation with Rocco Murano, publisher, FCL, Pawling, New York, July 5, 1989. This division in the structure of giving appears to have continued over time, although the evidence is less explicit and necessarily must be conjectured from the visibility of large donors and the publicity connected with their very large gifts periodically.

69. NVP to Paul Tillich, Apr. 22, 1960; Paul Tillich to NVP, Apr. 27, 1960; NVP Ms. Coll. Peale had noted on his letter an aside to send a copy to his son John, who was then studying with Tillich and was apparently fond of him.

70. John Sherrill, *Christian Life*, February 1970. The specific audience that was referred to here was the business community, but the comment applies equally to all

his audiences. Peale has taken pride in his ability to "adjust" his language to his audience. He frequently told of having to do that when speaking to the evangelical congregation at Ocean Grove, New Jersey, where they wanted to hear, he said, about "the Blood of the Lamb." He said he learned how to accommodate his language to their needs. Interview with NVP, July 5, 1989, Pawling, New York.

71. At the time I viewed the tapes in 1989 only four had been produced; of those, only one referred to Peale.

72. The former Lorimar executive, Mark Lambert, remained at PCI only a few years and then for reasons not clear returned to California.

73. Interview with Mark Lambert, July 7, 1989, Pawling, New York.

III

The Message

The impact and national celebrity of the movement Peale initiated was based more on its distinctive message than its popular messenger. The old/new message of Peale's practical Christianity was the common-sense conviction that the human mind, or Mind, was inherently good and trustworthy. A populist expression of middle America, it disputed the claims of both establishment leaders who argued for the special wisdom of those with gifts of breeding or intelligence, and those of fundamentalist believers who pressed for the primacy of a saved elect. Both views implied a privileged class: of genius and lineage in one case and sainthood in the other. Pealeism understood itself as an enemy to privilege wherever it was found, whether in academy, church, or legislature.

Peale's version of practical Christianity claimed high regard for the individual mind because it identified it as the vessel prepared for the Presence of God. It was imperative, he maintained, that the individual make an active and dynamic harmonial connection with this Presence, an act of conversion and commitment that in the context of Peale's ministry coincided with events of personal crisis. What was crucial to the implementation of personal empowerment was utilizing personal will, recognizing that one had the ability to alter life crises or to alter personal attitudes and behavior. The members of his invisible church were usually people with traditional ties to mainstream Protestantism, frequently members of churches with an evangelical background who found their local parishes failed to meet their needs. Peale gave voice to the discontent they described, often in communications to him but hesitated to express publicly.

That discontent was frequently framed in terms of criticism of an establishment and bureaucracy, in state as well as church. Within the church, the discontent was a critique of a liberal leadership derived from the legacy of the New Deal that had taken control of the spiritual homes of local worshippers, allegedly ignoring their needs in favor of more distant, seemingly altruistic, social programs. In the political realm, the populist charge was again directed against those liberals who seemingly destroyed the once-hegemonic values of American culture in the process of gaining bureaucratic control.

The influence of Pealeism fell along three lines. It nurtured a populist rhetoric of democratization that was ultimately absorbed by the new evangelicals. It suggested techniques for self-empowerment that enabled some supporters to become more effective dealing with personal crises, illness and pain, and behavioral or relationship issues. And it promoted a strongly individualistic message as a totalistic one, affecting all aspects of the culture, the political as well as the religious.

CHAPTER 6

Politics: Or "Ninety-Nine and Forty-Four One-Hundredths Percent" Practical Christianity 1935–1955

> Hence, the less government we have, the better,—the fewer laws, and the less confided power. The antidote to this abuse of formal Government, is, the influence of private character, the growth of the Individual; the appearance of the principal to supersede the proxy; the appearance of the wise man, of whom the existing government, is, it must be owned, but a shabby imitation.
>
> Emerson
> Essay on Politics
> c. 1844

In some ways Peale missed his calling. Despite extraordinary success in his chosen field, he was excited by the world of politics. The pull of things political was almost as strong for him as the call to the ministry. His youthful fantasy on visiting the Ohio statehouse, imagining himself occupying one of the positions of honor, was a portend of a future heated by the smoke and fire of the political arena. And yet it could also be said that he did not miss a special calling at all, that he simply transformed an early political drive to change society, to meld religion and secular culture, to the religious sphere. At least from the war years on, Peale's evolving message of populist religion was hinged to a conservative political agenda. Its full realization—important changes in the existing establishment structure—required strong political supports were it ever to be achieved. His populist religion and his conservative politics were, therefore, mutually reinforcing efforts, both relying on a constituent base of evangelical Protestant middle Americans. Their own postwar bid for acceptance, entitlement, and power—which has been termed an example of "bourgeoisification"—coincided with the fullest flowering of Peale's ministry, which they helped create.

Peale's Methodist parsonage childhood had given him an unshaded view of the rough-and-tumble contentiousness of local politics and left him with a lifetime appetite for political issues. Once he became a minister, he remained convinced that politics and religion were interdependent, their marriage consummated, if not at the time of the Reformation, then surely at the founding of America. His admiration of America's heroes and founders, of "great men," surfaced in his sermons.

He understood the minister's task to give life to the issues, in sermons surely, but also in whatever public forum presented itself. When a Marble Church member criticized him in 1956 for attacking candidate Adlai Stevenson in a Sunday morning service he responded, "As to my right as a preacher to discuss politics, I totally disagree with your idea that it has no place in the pulpit. It is my own policy to keep politics out of the pulpit, but I completely insist on not only the right but the duty of the preacher to deal with any subject which, in his judgement, should be discussed."[1] Politics would intrude on a sermon, crop up in a speech to a civic group, or appear straightforwardly in a press release. Certainly his very visible and active role in partisan politics until about 1955 was one of the defining qualities of his ministry and kept him in the continuing heat of controversy. Repeatedly challenged for his partisan activism, Peale repeatedly withdrew from the fray, only to be wooed back by another irresistible contest.

Experience and learning had persuaded him that politics and religion were mutually reinforcing; politics *was* religion; religion *had* a responsibility to the state. It was a vision of the complementary roles of church and state in keeping with a conservative evangelical model of the civil religion, similar in style to that once endorsed by the nineteenth-century evangelist Charles Finney. It was a view revealed in the way Peale represented politics to his congregation, and how, in turn, he interpreted their interests to those in the world outside. His religious calling, as he understood it, was a charge to the consciousness of habit-worn evangelicals, whose diligence, dependability, and inability to imagine new possibilities for themselves made them especially vulnerable to control by elites. His political calling, just as urgent, depended on reminding his congregation of their civic responsibility because of God's special providence for them as a people. He saw this mutuality best revealed in the golden age of "the city upon a hill" and now threatened because of human meddling and political mistakes. In another, more pragmatic sense, it meant monitoring the activities of liberals and checking them when possible.

Politics, Populism, and Prohibition

Peale first warmed to partisan politics through Prohibition. As a young man growing up in central Ohio, Prohibition was part of the religion of his family. It held the same place in his community and church, as indeed for

Protestants generally it served as a unifying "interest group issue." Straddling the realms of the sacred and the secular, Prohibition became a representation of the particular way its supporters viewed their culture. Its function as a defining interest group issue reinforces Robert Wuthnow's assumption about the role such issues play in a civil religion, namely, that

> both sides [religious conservatives and liberals] have a distinctive view of the larger society, including a civil religion that not only legitimates certain kinds of political issues but also provides their own organizations with a positive role to play in relation to the public interest.[2]

Prohibition became for people of Peale's generation a potential means for transforming the entire culture.

Although the issue itself failed, the campaign for its enactment remained for him an example of how a grass-roots political fire can affect national destiny. That legacy of youthful political excitement, like the similar sentiments of civil rights activists many years after "the Movement," helped shape his view of politics as, ideally, a local grass-roots, nonprofessional effort. As he got older and his move to Marble Collegiate Church required him to leave behind his ties to Methodism and consequently old denominational loyalties, Peale invested his new hopes for social change not in a single issue like Prohibition or a cause. He now pinned his hopes on "the Church," a transdenominational body that he and other Protestants identified as the agency for America's civil religion. Only this generic, militant Protestant force, sufficiently aroused, could restore the nation's seventeenth-century Anglo-European roots and harness nineteenth-century values to the gifts of the modern age. But by the 1940s, discouraged by what he felt to be the liberal leadership of an increasingly ecumenical mainstream "Church," Peale refocused his efforts on programs that had the flavor, if not the reality, of grass-roots campaigns. This was a redefined view of the civil religion that cast it outside the authority of institutional life, sustained instead by the grass-roots energy of popular culture and positive-thinking individuals. During the religious revival of the 1950s these two forces came together, as the political preferences of Cold Warriors joined with evangelical religious sympathies in a populist surge. When the revival waned, Peale shifted his hopes for political change to his Foundation for Christian Living.

His political philosophy, like his theology, was a hybrid product, consisting of high Toryism mixed with democratic populism. The Tory tendencies came originally from a youthful Methodism notion of heroism, from his admiration of Methodism's founder, whose achievements Peale credited with sparing England a "French-style revolution."[3] He imagined John Wesley a religiopolitical reformer, saving his country from bloody violence and lifting the working class to bourgeois status, an early example of *embourgeoisement.* There were also historical reminders closer to home, from the time of Charles Finney at least, which not only restated the connection between religious revivalism and political conservatism but, almost as important, made a practical point about the value of wealthy benefactors to a revival.

His Toryism and his populism both surfaced in his partisan activities over the years. Peale could as easily support the fiscal conservatism of wealthy men of affairs as he could inveigh against local Democratic politicians who condoned sleazy movies. And because the press gave generous coverage to his political views, he was constantly being sought out by individuals with political axes to grind. As a result, he often jumped on bandwagons whose real destinations he did not know. Eager to please, afraid to offend, lacking experienced confidants to give him advice, Peale repeatedly made poor choices. He stumbled into nasty nests of plotters, and stung by bitter criticism, vowed never to return, devoting himself "ninety-nine and forty-four one-hundredths percent of the time" to his ministerial calling. Part of the difficulty stemmed from an inability to distinguish boundaries and set limits. He was always trying to please an audience, whether of two people or two thousand, and he was repeatedly caught in compromised, contradictory situations. Yet he took more obvious risks politically than he did theologically or religiously, partly a measure of the importance of the issues as he saw them and partly the consequence of being a loner and not having a reference group to point out potential dangers. Politics was a game for the "man's man" he so admired, while religion too often seemed a parlor pursuit for the timid.

Even his critics conceded that the most egregious of his political follies were the result of foolish choices. Led into murky situations, Peale was taken advantage of, his name and reputation used to lend credibility to dubious schemes. But he was not led blindfolded into these activities, making it difficult to tell where bad politics ended and bad judgment picked up.

The four conservative groups he was involved with between 1935 and 1955 provided training ground for his own national ministry. During that twenty-year period Peale participated in the National Committee to Uphold Constitutional Government—in 1941 renamed the Committee for Constitutional Government—Spiritual Mobilization, the Christian Freedom Foundation, and briefly H. L. Hunt's Facts Forum. These organizations introduced him to mass mailing techniques and marketing strategies for securing potential backers and to means for enduring the pillorying of the press. Here as in his business contacts he came to know on close personal terms wealthy and powerful men. What he traded for these associations, and seemed not to comprehend, were the vestiges of goodwill from liberal professional colleagues and the opportunity to have his religious ideas taken seriously by them.

Until *The Power of Positive Thinking* appeared in 1952, Peale's political activities were largely responsible for the attention he received from the press and critics. The extent of this media exposure, which made him a public figure in the general culture, introduced him to a larger national audience. By the start of the war, with his national ministry just beginning to take off, his political activism gained in intensity.

Learning the Ropes: The Committee for Constitutional Government

It was hard to miss the political implications of a Peale sermon. When he said from Marble's pulpit in 1934 that there existed in the country "a tendency to encroach upon individual freedom. A sinister shadow is being thrown upon our liberties," few listeners missed its anti-New Deal meaning.[4] And when he advocated a revitalized Protestantism, complete with another Martin Luther, saying, "Protestantism is the only form of Christian religion adequate for the new day. . . . No religion of authority can meet the test of modern life. Protestantism is the religion of the spirit, not content to linger in ancient cloisters and hold fast to old shibboleths," the ethnic implications were easy to understand.[5] In 1936 he offered prayers for Republican hopeful Alf Landon, which was the same year he joined the National Committee to Uphold Constitutional Government.

During the thirties and even beyond Peale's basic political sympathies, as distinct from his partisan activities, were largely consistent with those of mainstream Protestantism. His anti-New Dealism was shared by most of his fellow clergy, many of whom, in a pre-Vatican II ethos conducive to mutual suspicion between Protestants and Roman Catholics, also participated in his ethnic bias. According to a 1936 *Literary Digest* poll of 21,606 clergymen, 70.2 percent, or 15,172, said they disapproved of the acts and policies of the New Deal.[6] Other evidence indicated that among church publications, only *Christian Century* supported Franklin D. Roosevelt in 1936, with most Protestants voting for Alf Landon that year. Protestants continued to oppose the New Deal, if by declining margins, as Thomas E. Dewey gained 37.5 percent of the Protestant vote in 1944 with 37 percent going to Roosevelt.[7] That clergy and congregations should be in concert on these issues is not surprising, since other evidence revealed that in churches without an episcopal tradition, particularly urban ones, the qualities preferred in a minister were those associated with the Rotary Club image suggested by Bruce Barton's *The Man Nobody Knows*. Search committees wanted a man with "personality" and business acumen, a good preacher who avoided controversial public issues in his sermons and unpopular political activities in his ministry.[8]

That Peale received only token criticism from his local congregation for his political views indicated that most members agreed with him and that his status conferred a considerable degree of freedom of expression. The press viewed him as something of a bellwether for changing sentiments and issues and usually made a good story out of any Peale sermon that had political connotations. In 1942 the New York *Sun* described him as "the decidedly urban type of cleric, handsome and impressive, filling his church with his weekly anti-New Deal philippics. He is possibly the outstanding public affairs cleric since the days of Bishop Cannon and the late John Roach Straton."

Although Peale might not have interpreted it that way, the paper meant the association to be flattering, for it went on to explain that Peale

> has been a vigorous assailant of the New Deal, preaching eloquent sermons against bureaucracy, official bungling, muddling and meddling, invasion of individual rights, wrecking of American traditions, coddling the unemployed, providing relief for the undeserving, knuckling to union labor, the menace of a third term, in fact, the entire category of New Deal sins as he sees them.[9]

This sort of continuing attention by the press increased the likelihood that Peale would be sought out by political interest groups.

And in the winter of 1936 Peale was sought out. At that time Rochester, New York, publisher Frank Gannett approached him with a request to become a founding member of the National Committee to Uphold Constitutional Government. As Gannett explained, the committee's immediate target was the President's plan to increase the size of the Supreme Court, the so-called Court-packing effort. Like Peale, the other organizing members of the committee were also well-known individuals, with allegiances ranging from moderate to conservative and including at least one credentialed liberal, John Haynes Holmes—who quit the group after a year. Roosevelt's strategy, fueled by his massive win over Landon and designed to break the conservative majority on the Court that was holding back economic legislation, had touched sensitive nerves and therefore would ultimately fail due to popular opposition. In the already heated climate of opinion, demagoguery flourished, and charismatic speakers like Father Coughlin, Gerald L. K. Smith, and followers of the late Huey Long could sway large audiences, who inevitably found points of identification in the diverse presentation of anti-Roosevelt platforms.[10] By comparison, Gannett's group, professionals from law, medicine, academia, and the church, appeared as a respectable citizens' organization formed in response to a specific issue. In fact, none of the politically oriented organizations Peale joined—the Committee for Constitutional Government, Spiritual Mobilization, the Christian Freedom Foundation—claimed it was political, but rather that its interests were educational and religious.

Gannett's committee was actually formed from the remnant of a highly partisan group he had assembled the previous summer to support progressive-isolationist Senator William Borah of Idaho for the Republican presidential nomination. The owner of a string of twenty-one newspapers, with important connections, including a seat on Cornell University's board of trustees, Gannett was a self-professed rags-to-riches success, then at the peak of his career. With himself as chair, Gannett set up committee offices in New York City and installed controversial newspaperman Edward A. Rumely as executive secretary.

Gannett provided the seed money, and the committee began a massive mail distribution campaign to point out the "unconstitutionality" of Roosevelt's Court-packing proposal. By its own calculations, the committee sent 10 million envelopes to "leadership individuals," some of whom became subscribers, making reportedly "small" contributions.[11] When the Court plan was defeated, the committee felt rewarded. Its effectiveness was attributable to a number of important factors, including its ready access to the news media, its possession of valuable mailing lists, and its ability to raise money to fund huge mailings, including telegrams. Equally important was Edward Rumely's skill as an organizer and strategist, a behind-the-scenes power in the committee and allied organizations for the next fifteen years.

Critics of the committee were especially wary of Rumely, whom they described as a convicted felon with a record as a Nazi sympathizer. Peale and other Rumely friends argued that Rumely was a victim rather than a victimizer, a man wrongly accused and then pardoned. Nevertheless, he brought a blemished record to his work on the committee. The relevant facts in his history were far from clear. Descended from German grandparents and educated abroad, Rumely had purchased the New York *Evening Mail* in 1916 with funds borrowed from two Americans then residing in Germany. Charged by the government in 1918 for failure to report his German loan, Rumely was accused of plotting to propagandize for the Germans and was convicted and sentenced to a year and a day in prison. Rumely's defense was that wartime conditions prevented his reporting the loan at the time. He was granted a full pardon by President Calvin Coolidge in 1925 and was released from jail after thirty days. His accusers said Rumely's pardon resulted from the intervention of his friend Henry Ford.[12] The committee realized that Rumely's past was a liability and kept his name off its correspondence while still encouraging him to direct its efforts.

On the heels of the Court-packing fight, the committee geared up for another round, this time against Roosevelt's plan to "purge" the party of southern Democrats who voted with conservative Republicans and to challenge Roosevelt's bid for a third term. Peale had particularly strong feelings on the no-third-term issue. In March 1938 Rumely and the committee's treasurer, Sumner Gerard, were called to appear before the Senate Committee Investigating Lobbying Activities to be questioned—unsuccessfully—about the sources of the group's finances; they were dismissed subsequently with a warning about continued Congressional scrutiny.[13]

By the fall of 1940, Gannett's replacement as the committee's chair, ex-Congressman Samuel Pettengill, was back in Washington to answer more questions about committee finances as well as about possible links—conspiratorial or otherwise—to other organizations. Examined by Guy Gillette, the chair of a Special Senate Committee Investigating Campaign Expenditures, Pettengill was also pushed on the accuracy of the group's designation as an educational agency. A member of the Senate Committee pointed out that the organization's ads had appeared in many newspapers under the heading "Political Advertisement" and that its literature showed up at campaign functions for presidential aspirant Wendell Wilkie. In response, Pettengill stated that the 15 million pieces of mail sent out opposing the "purge" constituted a massive educational endeavor, as did his own articles and book on the subject, *Jefferson the Forgotten Man*.

As if to give substance to his claims, Pettengill offered in evidence a letter written by Peale, then serving as secretary to the Gannett group. Read into the record, Peale's statement was essentially a testimonial to the integrity of the Committee to Uphold Constitutional Government, his own reputation presumably serving as the consideration. To answer the question about income, he noted without explanation that in the group's three-year history, it had received contributions from 35,000 individuals in amounts "averaging not over $40 per contributor during the three years." Further, Peale explained that he was himself part of the committee's educational program, having been the principal speaker the previous evening, October 22, 1940, at a Carnegie Hall rally to outline the need for a Constitutional amendment against a presidential third term. Peale identified himself as the author of a no-third-term letter to all members of the clergy, sent through the auspices of the Committee and under his personal signature.[14] If the Senate Committee, worried about questions of linkage, was concerned about the Gannett group's links to other agencies, then, Peale was affirming, he was the link, and the tie was to the clergy. His letter made clear that he was the group's most respected representative to the general public.

Spiritual Mobilization

Peale was correct in noting that it was he who provided the link between the Committee to Uphold Constitutional Government and the clergy. Although presumably for the benefit of the conspiratorially minded, he described the connection not as a link but as an aspect of his professional friendship with other ministers. And at the same time that Peale was most involved with the committee, in the early 1940s, he was also giving time and attention to Spiritual Mobilization, a politically partisan group that claimed open ties to the religious community.

Peale had cultivated a long-distance friendship with West Coast minister James Fifield, who in 1934 founded the conservative Spiritual Mobilization. In addition to working tirelessly for Spiritual Mobilization, the privately wealthy Fifield served the First Congregational Church of Los Angeles, reportedly the largest church in the denomination. Engaged in a fight against "pagan stateism"—"fight" being an operative word in the literature of anticommunism—Fifield had formed Spiritual Mobilization with a distinguished advisory board of conservatives: college presidents and professors, scientists, physicians, and other professionals.[15] By 1944, Peale's name was among them. Fifield was an articulate speaker, whose sense of urgency was clearly evidenced in his popular 1947 sermon, "The Cross vs. the Sickle," denouncing the international postwar agreements arranged by Presidents Roosevelt and Truman:

> If the American negotiators had been in the employ of Russia, they could hardly have served her better! And we still don't know it all! . . .
>
> So we ought to denounce all agreements, secret or public, in which our country has sold out any of its principles for a mess of the pottage of expediency. And we ought to return to those policies and principles which have made America great and strong and honorable and good so that we love her. Too many things have been too long hushed up. This country through her leadership has sinned and erred and has got to make atonement. Her first step is to confess her sins, the sins at Yalta and Teheran and San Francisco and so forth. . . .
>
> Well, all I have been trying to say this morning is that it's time to quit temporizing with the sickle and get back with renewed faith, hope, and devotion to the Cross and all for which it stands.[16]

Skilled as a fund raiser, popular at Rotary Club and Union League luncheons, Fifield aimed to get secure commitments to Spiritual Mobilization from the 10,000 ministers regularly circularized by its mailings. One strategy that he hoped might help him reach that goal was a "pagan stateism" announcement, including a full list of the advisory committee, placed in various organs of the religious press. That announcement warned in part: "For a decade America has been moving toward pagan stateism of the sort which makes citizens servants instead of masters of the state. . . . The clergy of America have fundamental responsibility to stand watch over . . . Christian ideals, presently in peril but not yet beyond saving." One of these advertisements showed up in the August 1944 issue of the *Defender*, a nasty, anti-Semitic, anti-Communist publication of fundamentalist Baptist evangelist Gerold B. Winrod.[17] When the "pagan stateism" notice appeared, not in the *Defender*, where they probably missed it, but in *Christian Century*, a group of five liberal ministers and professors wrote to Peale to ask him to explain "ambiguous" statements, which seemed "to conceal, perhaps, certain meanings" about a movement they believed was having "somewhat extraordinary significance."[18] Peale's response was that he did not have the time to answer the questions they raised.[19] Shortly thereafter, *Christian Century* refused to accept any further ads from Spiritual Mobilization.

During the early forties, Peale was Spiritual Mobilization's representative in New York City, a regional agent for what Fifield envisioned as a national grass-roots program to "mobilize" the clergy. In both structure and intent, the organization resembled a "special purpose group," the kind of cross-denominational interest group that students of the phenomenon say began proliferating after 1960, frequently in response to expansion of the state's role.[20] In 1946 Spiritual Mobilization claimed 7,327 regional representatives, with the greatest number, 555, in Pennsylvania, followed by California with 419, Ohio with 402, Illinois with 386, and New York with 384.[21] Regional meetings were supposed to bring supporters together, but the organization was essentially an impersonal communications—many called it propaganda—network, held together by Fifield's drive, energy, and writings. His fear was collectivism, and he trained his fire on anything that seemed to him to suggest evidence of it: schools and colleges, labor unions, the alcohol trade, and the Federal Council of Churches, this last blast so alarming to Peale that he asked his friend to muffle it.[22] According to columnist Victor Riesel, Spiritual Mobilization was able to fund its activities through contributions from industrialists, such as a $50,000 gift from the National Association of Manufacturers, and through the distribution of its literature, including the book *The Spirit of Enterprise*, written by the anti-Wilkie isolationist Republican Edgar M. Queeny.[23]

As a result of their cooperative efforts in Spiritual Mobilization, the Peales—Ruth and Norman—and the Fifields—Helen and James—became family friends as well as colaborers in the Cold War vineyard. The Fifields owned an extensive cattle ranch outside Los Angeles, with Mrs. Fifield presumably the source of the family's wealth. The Peales visited with the Fifields on their trips to the West Coast, and the two Fifield daughters became friendly with the Peale children. Yet Peale was sufficiently sensitive to the controversial nature of Fifield's endeavor that he evaded discussion of the subject when it was possible. On one occasion when a Pittsburgh Presbyterian clergyman questioned him about the group's source of funding, Peale answered, "I am not very well acquainted with the details of operation of Spiritual Mobilization, and have no knowledge of the amount of contributions."[24] When he was questioned by another clergyman known to be sympathetic to the organization, however, Peale was able to be less guarded and remarked that "Dr. Fifield is one of the most amazing personalities I have ever known, and has a strong support, both financial and otherwise, behind the movement."[25]

Popular with conservatives in business and the church, Spiritual Mobilization was condemned by liberals. The *Christian Century* decision to refuse its notices reflected the sentiments of theologian Wilhelm Pauck, who called it the "most dishonest spiritual movement of the day."[26]

The Christian Freedom Foundation

In 1948, his energy and health flagging, Fifield approached Peale with a proposal that he become the New York director of a reorganized Spiritual Mobilization, once a significant part of its operation was transferred there. Peale hesitated. Fifield assured him that any planning sessions for the transition would be "friendly" and without "difficulty." To pay for this new part of the program, Fifield went to the friend and benefactor of conservative causes, Presbyterian layman J. Howard Pew. Pew's view of the appropriate role of the church in political activities had been a matter of record for some years, ever since he and other like-minded lay people had joined their efforts to fight the liberal pronouncements of denominational executives and ecumenical leaders. They had been encouraged in their own struggle by the assertive challenge to social gospel measures launched back in 1936 by the Layman's Religious Movement within the Methodist Church. After consultations with Pew, Fifield and Peale decided that the base of Spiritual Mobilization would remain in California, but a part of the operation, complete with mailing lists and personnel, would be transferred to New York, where it would be renamed the Christian Freedom Foundation, and would produce a publication to be known as *Christian Economics.*[27] Pew, who had been generously underwriting the rent for Spiritual Mobilization offices around the country, was the primary backer of the project, a matter that concerned Peale and Fifield because of the potential for bad publicity.[28] As Peale said to Howard Kershner, the new editor of *Christian Economics,*

> [Dr. Fifield] feels that it is essential to the success of *Christian Freedom Foundation, Inc.*, that other financing besides that of Mr. Pew's contribution immediately be obtained. He says that if it becomes known that this is financed entirely by Mr. Pew that it will bring heavy attack upon it.[29]

Peale therefore became the new chairman of the Christian Freedom Foundation, a job he combined with an increasingly hectic ministerial schedule and added on to his responsibilities to an active young family. Fifield had foreseen the difficulties of the combined demands and suggested to Peale that the compensation from his new appointment could replace his lecture work on the road.[30] Ignoring Fifield's suggestion, Peale did not hold the top job long; elected chairman at the organizational meeting of the group on April 17, 1950, he resigned by May 9.[31] He returned to the group as treasurer in 1962, after the Foundation agreed to two conditions—that Peale be a "real" and not "rubber stamp" treasurer and that his appointment have the "whole-hearted approval of Mr. Pew."[32] He stayed on the periphery of the organization, although his opinions still carried some weight with its officers, for his objections to heavy criticism of the United Nations led the group to tone down some of its expressions.[33] But he remained in basic sympathy with the values of the organization, as was evidenced by his comments on a statement by theologian Robert McAfee Brown that appeared in the liberal journal *Christianity and Crisis.* Brown had said that he doubted the organization's fundamental "proposition that "freedom and Christianity" are interdependent." Peale thought Brown's comment reflected the fact that *Christianity and Crisis* "lean[s] rather heavily towards Communism."[34]

Although Peale decided not to assume the leadership of the Christian Freedom Foundation, he continued to keep alive his ties to Spiritual Mobilization through his friendship with Fifield. It had been at Fifield's request that he served on its advisory board and as its New York field representative, and that he had taken to the lecture circuit on its behalf. During the summer of 1951, as he contemplated suitable locations for drafting *The Power of Positive Thinking*, Peale accepted Fifield's offer to take on a speaking tour of Hawaii. Fifield was agitated by the labor situation on the California docks and Harry Bridges' role in it, fearing that the West Coast problem was in danger of being exported to Hawaii. His friend's visit was part of his plan to avert labor problems in the islands. Peale explained to an acquaintance that he was "speaking in Honolulu under the auspices of Spiritual Mobilization, a national organization of which Dr. Fifield is the head, and he has engaged me for these addresses."[35]

Peale's visit to Hawaii was meant to focus on the local labor situation. Both Spiritual Mobilization and the Committee for Consitutional Government supported right-to-work laws, which banned union shops. Peale found a local labor environment that fulfilled Fifield's expectations, one he described as "sinister," because of the efforts exerted by a "communistic leadership" attempting to control the rank and file of workers.[36] During his trip, Peale met with the heads of the Transportation System and the Employers Association and breakfasted with the head of Dole Hawaiian Pineapple Company.[37] He spoke before an overflow crowd at a Chamber of Commerce meeting that was recorded and later broadcast on local radio. He went out to communities and talked with ministers. Encouraged by his reception, Peale wrote to Fifield from the scene, "If we can get these people to be aggressive and to equal in aggressiveness that of the IOWU, it is my judgment that the common people over here, the workers, will respond. The workers are a fine type; at least those that I have seen."[38] He told a local clergyman worried about how to deal with the situation that what was necessary "to change the communistic people . . . [was] to bring everybody into harmony with the spirit and ideals of Christ."[39]

Peale's public position on unions was that it wasn't unionism per se that he objected to but rather the coercive and communistic nature of some unions. He pointed to his personal record of support for the textile workers in their struggle with management when he was a student minister in Berkeley, Rhode Island. He idealized a kind of frontier model of the self-employed, claiming that "even if they [labor leaders] all were reincarnations of Abraham Lincoln, nobody has the right to force me, or any other American, to join any organization to get a job."[40] Certainly during the forties and fifties he found it easy to make a simple identification between "labor unions" and "Communistic influence," convinced in 1950 that "there is a strong Communistic impulse behind the labor union movement."[41]

There is no indication that Peale's visit to Hawaii had any impact on local labor conditions in the islands. Nor did critics of Spiritual Mobilization even seem to be aware of his trip, although it was publicized in the islands.

Under Cover

Peale did not fare as well when it came to the wrath of the critics who monitored the work of the Committee for Constitutional Government. He was more closely connected to the Constitutional Government Committee than he was to Spiritual Mobilization—in part because the committee was headquartered in New York—and when its activities were held up for review by government watchdogs and diligent opponents, his name was revealed as one of those caught in the web of conspiratorial politics. But he had joined the Constitutional Committees without hesitation, and between 1942 and 1945, he served as chairman of the Committee for Constitutional Government, resigning when he recognized the negative effect this role was having on his efforts to develop *Guideposts.*

Liberal critics who kept these ultraconservative organizations under surveillance, like the editor of *PM*, George Seldes, and his Friends of Democracy, searched for conspiratorial links among them. Seldes made an explicit charge that

> This time the connecting link is the Rev. Norman Vincent Peale, pastor of the Marble Collegiate Church, who is on the advisory board of Spiritual Mobilization and who is also chairman of the Gannett-Rumely Committee. Peale once gave the invocation at a meeting of super "patriots" where Elizabeth Dilling and Edward Lodge Curran were the chief speakers.[42]

Peale said that the "sinister implication" of a "link" was "absurd," that Fifield was his friend.

Seldes and the pseudonymous author John Roy Carlson put Peale on the defensive about the 1943 Dilling-Curran meeting at the Hotel Commodore. Carlson's book, *Under Cover*, claiming to be an exposé of "the Nazi Underworld of America," was written when Carlson went under cover to secure his evidence. Carlson resurrected an event Peale hoped had been buried when he wrote:

> Rumely is boss of the Committee for Constitutional Government and second in command to Frank E. Gannett, publisher of a string of newspapers and founder of the Committee in 1937. As soon as the Senatorial investigation was over, Rumely literally went underground and erased his name from the Committee stationary. But he continued to run it by appointing a docile Protestant clergyman as "acting chairman and secretary" who visited the office only occasionally. He was the Reverend Norman Vincent Peale, once a joint speaker with Mrs. Elizabeth Dilling and the Reverend Edward Lodge Curran at a "pro-American mass meeting sponsored by more than 50 patriotic organizations" at the Hotel Commodore in New York.[43]

At the bottom of the page on which the text appeared was a reproduction, in large black lettering, of the part of the Commodore program that listed the invocation, followed by Peale's name and his title as minister of the Marble Collegiate Church. Peale consulted a lawyer in an effort to file grievances against Carlson and his publisher, E. P. Dutton, for libel, but was advised he lacked sufficient evidence. He had given the invocation, and Carlson had noted that fact. His reputation had been badly tarnished by seeming to associate with Mrs. Dilling, a person the federal government ranked among the worst hate mongers in an environment colored by hate; she was a "patriot" who smeared liberals, Jews, African Americans, and other ethnic groups with the same broad brush.

Peale's version of the Dilling-Curran event differed markedly in interpretation. He claimed that a woman in his church kept badgering him to speak to an organization to which she belonged, and he resisted, he said, because "I do not like that kind of patriotism." When she finally suggested that he might offer an invocation if he could not make a speech, he relented, explaining that an invocation was a prayer and that he would agree to pray anywhere, in "prisons, saloons, political conventions, labor union rallies," and even "in Hell." He stayed at the Commodore, he recalled, no more than ten or fifteen minutes, giving the invocation and then departing immediately. In an angry letter to the president of E. P. Dutton, he charged that the publisher had indicated his integrity by answering he "did not know" when asked about Carlson's implication that he, Peale, was "a member of the Nazi underworld of America."[44] In a letter to Carlson, Peale admitted that "Undoubtedly it was a mistake for me to be at the meeting," but he charged the author with deliberately misinterpreting the circumstances to discredit the committee and Peale himself.[45]

Carlson's work, a long, colorfully written exposé of right wing hate groups and protofascists, was favorably reviewed by *The New York Times* critic, Orville Prescott, who said the book was "of sensational importance to every man and woman of good-will" who believed that "the barbarians within our borders must be defeated as well as those outside."[46] Peale's chief accusers, Carlson and Seldes—the latter many in the Committee believed to be the real power "behind" Carlson's work—released him from the company of the "barbarians" only to indict him for being a dupe. In *The Plotters*, his 1946 sequel to *Under Cover*, Carlson referred to a long and moving explanatory letter he said he received from Peale regarding the Commodore meeting, which Carlson said "I believe." But in his opinion, said Carlson, Peale was wrong "to allow his church to be used by a high-powered group for political ends. Rev. Peale's name dignified its work with apparent religious sanction."[47] George Seldes was of a similar mind. In a letter to Peale, he said he withdrew "any charge of your being a native American fascist," although that left him with the conclusion that Peale knew nothing about the organization he worked for and that he was "the innocent tool of the worst outfit of native American fascists now trying to undermine the Constitution and the American way of life."[48]

As the story unfolded, Peale and the committee were surrounded by a chorus of protests from labor organizations and Jewish groups, the disapprobation of his New York neighbor Rabbi Stephen Wise being particularly wounding to Peale.[49] The real target for most of the criticism was Edward Rumely, who was perceived as the major player in the committee, allied with professional conservative politicians like Pettengill, New Jersey Senator Albert Hawkes, New York Congressman Ralph Gwinn, and others in the northern Republican-southern Democratic reactionary alliance. Carlson even excused Gannett, saying that he had been "victimized by political friends who have sought to use him to their advantage." The difficulty, said Carlson, was that "Gannett and his Committee are symbols of an era which made America strong and great," but they continue to search in the "blind alleys of yesterday," where they find "no hope, nor encouragement, no objective enlightenment" for the future.[50]

Peale made a costly error in judgment in consenting to the invitation of the wealthy, well-placed woman in his church to appear at the Hotel Commodore. It was another instance of his need to please and be accepted, to be the kind of person he would later describe in *The Power of Positive Thinking* as a comfortable "old-shoe, old-hat kind of person." It would not be the last time he would be drubbed by the press for his political mistakes.

Personal Politics

Until he resigned his role in H. L. Hunt's Facts Forum in 1953, Peale had only occasional misgivings about putting himself in vulnerable, vanguard positions in conservative politics. What misgivings he did have usually related to support for *Guideposts* magazine. While he ran some risks, chiefly from the scrutiny and attendant public reaction of liberal critics, he believed that they were worth taking and did not seriously jeopardize his personal ministry. There were, in fact, ways in which his ministry was enhanced, as wealthy benefactors opened their checkbooks to him in response to a particular stand he had taken. Eventually, however, Peale was faced with a choice, or rather continuing choices, over whether to remain engaged with a controversial conservative issue and risk the approval of members of his national audience or to detach from partisan activity. Although he remained energized by a passion for politics, he learned that to continue as a public partisan for unpopular, or controversial, issues took a toll on his personal ministry. After withdrawing from Facts Forum in 1953, Peale maintained a low-key political profile until he re-emerged at the time of the John F. Kennedy presidential campaign in 1960.

During the heyday of first his anti-New Deal and then Cold Warrior days, he expressed no reservations about the wisdom of being outspoken on the issues. When Peale gave his "no-third-term" speech in Carnegie Hall in

October 1940, he told his father that despite the criticism he felt persuaded by his convictions that it was a "patriotic duty and I am in for it."[51] As he saw it, clergymen could have "traditional, conservative" ministries while still holding "an exposed position in support of their convictions," as he tried to do in his own work."[52] Zealous, close to the issues, he offered his opinion on the full gamut of conservative causes. Because he was so often on the lecture circuit, he was constantly facing new opportunities to "lambaste," as he said, the New Deal.[53] In 1941 he proposed a Protestant version of the Catholic Index to combat "profanity and immorality" in literature, although at the time he could think of only four books to recommend as desirable reading.[54]

Some of his politicking was personal and local, generated by requests from friends and church members. When E. F. Hutton, Peale's friend through the activities of the Constitutional Committee, worried about the future of the right-to-work law, Peale reassured him that as a result of his far-reaching lecture commitments he "followed the theme of the right to work, many places."[55] He warily sidestepped the concerns of a church member over the admission of 100,000 Jewish refugees to the United States, offering a warning against feelings of anti-Semitism while managing a promise to "say something about this matter" of the refugees to his Congressional representative "or otherwise."[56] On his own home ground, he protested to New York City's Mayor Fiorello LaGuardia about "indecent" theater productions and about the starting time for Sunday night blackouts, which left people departing his evening service in complete darkness. He alerted the Committee on Constitutional Government to the labor union strategy of using comic books to circulate its message and suggested that the committee might want to follow suit. Within his own church, he distributed copies of Pettengill's book and committee-sponsored editions of the Bill of Rights and the Constitution, first to Sunday School teachers, then to youth groups, and then wholesale to members of the congregation.

Even in these personal political efforts, Edward Rumely stood ready to be a Peale mentor and assistant. He had a facility for remaining in touch with the busy cleric, always on the go, when others were often frustrated in the attempt. He clearly understood the value of Peale's image with the public. In the summer of 1941, as the Committee for Constitutional Government faced a shortfall in funds because only 4,100 of the 41,000 members had contributed, Rumely asked Peale, not yet chairman of the group, if he would write a "moving, gripping" letter on his church stationery to potential supporters.[57] When Rumely told Peale about rumors of left-wing efforts to organize southern blacks and asked him what might be an appropriate countermove, Peale answered with a suggestion for a meeting of the "most sensible and finer spirits among the colored ministers" to balance the charged message of Adam Clayton Powell.[58] On another occasion, inspired by the success of an essay-writing contest sponsored by the committee in

Texas public schools, Rumely approached Peale with a request to try something similar in the Marble Church Sunday School, as well as in the public schools in Pawling.

With Friends Like These

Between 1944 and 1946, the Committee for Constitutional Government was steadily embroiled in conflict with the Congress, either with Congressional committees or with Democratic Congressmen. Although Peale chaired the Constitutional Committee for part of the time—until 1945—he was somehow exempted from participation in the hearings, with Rumely the chief target of the investigations. Rumely's work with the committee had been under a cloud since he began, his increased scrutiny by congressional investigators a measure of both his effectiveness as a lobbyist and the heated Cold War political climate. Congress focused on how the Committee for Constitutional Government raised its money. Presumably with the support of his conservative friends, Rumely agreed to be the target for the federal probers.

Beginning in the spring of 1944, Texas New Deal Congressman Wright Patman launched an effort to get the committee to divulge the source of its revenue. Rumely confided to friends that Patman's real problem was that the committee's essay-writing contest in Texas was starting to cause "difficulty" for the congressman. Calling the committee "the outstanding Fascist group in America," and Rumely its "brains," Patman said his goal was to block the committee's lobbying efforts for a Twenty-second Amendment that would set a maximum tax limit of 25 percent on incomes. Patman pointed out that the committee had already achieved remarkable success with sixteen state legislatures, and were their proposal to be realized, said the Congressman, America would become" a paradise for millionaires."

As the fall elections approached, Rumely, Gannett, and Sumner Gerard—though not Peale, the chairman of the group—were back in Washington to appear before the House Committee on Campaign Expenditures. Asked about the identity of their contributors, Rumely stonewalled, as he had promised his friends at home he would do, in defense of "legitimate free speech." And when questioned about the possibility that committee funds were being diverted to Texas to defeat New Deal candidates, Rumely explained that they were Texas-raised contributions that appeared on the records in the New York City office for bookkeeping reasons only.

Back in his New York office, Rumely was visited by members of the Congressional probe, led by New Deal Congressman Clinton Anderson, and asked again to produce the list of contributors, Rumely once more refused. He failed to comply with a congressional subpoena of the records; indicted for contempt, Rumely secured a dismissal when the jury could not reach an agreement.[59] Tried a second time, in April 1946, Rumely was acquitted. His defense lawyer for the case was Neil Burkinshaw, whose services, critics alleged, were subsidized by Hearst interests.[60]

Rumely's trials revealed just how diligent he had been as a lobbyist. The investigation testified to the committee's public relations effectiveness, in an accounting which showed that in the years between 1937 and 1944, at a cost of $10 million, the committee had distributed: 82 million pieces of literature—books, pamphlets, reprints of editorials and articles, letters, and 760,000 books; 10,000 transcriptions, carrying fifteen-minute radio talks; 350,000 telegrams to individuals; and thousands of releases to daily and weekly newspapers, as well as full-page advertisements in 536 different newspapers with a combined circulation of 20 million.[61] The volume of evidence persuaded the House Committee to withdraw the organization's tax-exempt educational status, claiming that it was clearly a partisan group. Shortly thereafter the committee created an "educational" offspring, its offices adjacent to the committee's own, and its appointed sales manager Rumely's daughter.[62]

Rumely's troubles with the government did not end there. He continued to be questioned about the sales of committee-sponsored books, an obvious device for raising funds. The committee promoted a number of books—Pettengill's *Jefferson the Forgotten Man*; Thomas J. Norton's *The Constitution of the United States: Its Sources and Its Application*; Queeny's *Spirit of Enterprise*; and later John T. Flynn's *The Road Ahead*. The books were sent to leaders in industry, who were then expected to buy copies for their employees at inflated prices. A book that cost the committee 20 cents was sold to businessmen for 50 cents and upwards to a dollar.[63] Rumely was investigated by the federal Department of Justice for circulating Flynn's work, allegedly now at cut rates, and according to columnist and Rumely supporter Westbrook Pegler, Rumely was tried, convicted, and sentenced to prison, with the sentence subsequently being remitted because of Rumely's advanced age.[64]

At the time of Rumely's last encounter with the government, Peale was still active in the committee, although he had long since relinquished the chairman's role. He had written to Rumely in 1945 to explain why he needed to resign:

> I am sending you herewith a letter which I certainly feel badly to write to you. There is no alternative. It is a fact that my work with the Committee for Constitutional Government is hurting me in my own work. I cannot continue. . . . If I continue any longer without a definite statement that I am no longer connected with the Committee officially I might as well drop *Guideposts*.[65]

The continued publicity of the federal investigation of the Committee for Constitutional Government, at a time when Peale was serving as its chairman, brought him face to face with a major conflict of interest. The infant *Guideposts* project had come to claim his larger loyalties, and his very public and controversial role with the committee was costing him the good will of those whose public relations and financial help he needed.

The following year Peale delivered another swan song to politics. Writing in *The New York Post*, he said he had "arrived at the conviction that nobody can take any part in public affairs" without being subjected to smears. He ended on a sardonic note: "Therefore I have decided to give my time to trying to help people individually, and leave the creation of the brave new world to those who are wiser than I."[66]

In spite of his very visible ties to conservative interest groups, as well as his consistent opposition to New Deal programs, Peale not infrequently identified himself as a political independent, and even wrote to a friend, "I am not a conservative."[67] In fact, he had made some surprising concessions to liberal and/or bipartisan interests. He had participated in the National Conference of Christians and Jews, signed an anti-Franco statement from the American Committee for Spanish Freedom, and, with continuing misgivings, endorsed the new National Council of Churches when it took shape in 1950.[68] That both his wife and J. Howard Pew were to have roles in the National Council of Churches temporarily eased some of his doubts about the ecumenical agency.

Facts Forum

Even after transferring the creation of a "brave new world" to others, Peale remained politically vulnerable. In a dance of reciprocity, those with political causes to advance sought him for his public relations value, just as he recognized in them potential patrons for one of his projects. For two years, between 1951 and 1953, he participated in such a scenario as he was courted by H. L. Hunt's Facts Forum.

The "educational" organization sponsored by Texas oilman Hunt sounded extraordinarily catholic in its purpose: Facts Forum was described as a "nonprofit national educational organization devoted to the study of political science, soil and water conservation, and the art of living." Working from a Dallas base, it projected a national network of grass-roots meetings, "neighborhood discussion groups," which, after airing both sides of controversial issues, would ensure that a "clear majority will reach the correct conclusions"[69] In addition to these neighborhood meetings, Facts Forum planned a half-hour television broadcast from Washington, as well as shorter television and radio programs conducted by Facts Forum commentator Dan Smoot. Those who examined its literature were struck by its right-wing orientation and its opposition to the United Nations, particularly UNESCO.

Peale first met Hunt in October 1951 when he was invited to address a Texas-sized gathering of 50,000 at the Texas State Fair in the Cotton Bowl. Local friends of the two men introduced them by having them swap copies of each other's publications; Hunt was given samples of *Guideposts*, and Peale received copies of Facts Forum literature.[70] They continued to correspond, and by December Hunt was suggesting to Peale that he might want to call on General Douglas MacArthur, the former supreme commander of United Nations forces in Korea who had been fired by President Harry Truman and was then living in New York City.

Conservatives had begun to consider MacArthur as a viable presidential candidate in the upcoming Republican national convention, the potential brokered winner in what was expected to be a fierce struggle between Senator Robert A. Taft of Ohio, who would be acceptable to them, and General Dwight D. Eisenhower, who certainly would not. Although many Americans admired MacArthur for his military courage, conservatives were particularly heartened by his willingness to confront Truman and by his intransigent commitment to a showdown war in the Far East. It was with the hope of persuading the general to accept a draft that Hunt asked Peale to visit him, assuming that a persuasive argument from a popular and well-connected preacher would have a telling effect.

Peale accepted Hunt's suggestion and visited General MacArthur his Waldorf Towers suite on December 17.[71] The results of their meeting were inconclusive, although Hunt remained convinced that diligent lobbying efforts would get MacArthur nominated and began making plans for a Demand MacArthur Club complete with campaign buttons.[72] Subsequently Peale wrote an introduction to a laudatory pamphlet on MacArthur, which was circulated by, among others, conservative politicians to their constituents. Some recipients saw it as a thinly veiled effort to promote Taft's candidacy.[73] While that was not its stated intention, it would have been difficult for so loyal a son of Ohio as Peale to promote efforts that would deliberately sabotage Taft's campaign. Yet after his visit with the general, Peale assured another MacArthur booster, Frank Gannett, that "I would like very much to see him become President, and I am hoping and praying that such an event will take place."[74]

Hunt and Peale stayed in contact, through phone calls, letters, two or three personal visits, a Christmas gift of pecans from Hunt to the Peales, followed by a contribution of $500 to the Peales' upcoming television program.[75] In February 1952 Peale agreed to become one of nine "counselors" on the Facts Forum board, adding his name to those of the other undeniable conservatives already serving: General Robert E. Wood, W. G. Vollmer, John Wayne, Lloyd E. Skinner, Governor Allan Shivers, Dr. Francis P. Gaines, General Albert C. Wedemeyer, and General Hanford MacNider.[76]

Facts Forum neighborhood gatherings began appearing in the New York metropolitan area, and Marble Collegiate Church offered its chapel as one of the meeting places. A member of Marble's congregation who attended several of the sessions felt compelled to write to Peale to express her shock at the one-sided way they were conducted. She was deeply concerned, she said, that there had been no opportunity to contest the anti-United Nations argument of the Facts Forum representatives or to debate the anti-Semitic address of a man "so much identified with Nazi-like propaganda."[77] For the time being, Peale was willing to defend Facts Forum, to explain the controversial speaker as either a Fascist or a Communist plant, but he was beginning to understand that continued association with the organization could be unwise.[78]

After doing some homework on the organization, Peale came to his own conclusion about it. In November 1953 he resigned, telling Hunt that while Facts Forum was "conceived of as an objective and unbiased project," he had come to realize that it does deal "in the area of strong controversy." Still interested in maintaining their personal friendship, he observed to Hunt,

> One thing has been impressed upon me by those with whom I am associated in the church work and in my lecturing and book publishing interests; and that is that my influence as a spiritual teacher and guide suffers from activities and endorsements in connection with projects in the social or political areas of thought.

And so he said farewell to politics again. He had given the matter much thought, he wrote Hunt, and he had decided that "it is my function in life to give what time and strength I have to matters of a purely spiritual ministry."[79]

And that is essentially what he did according to his own design, with the notable exception of his foray into the 1960 presidential campaign. Peale lent his name to political causes he supported and was a kind of unofficial chaplain to Richard M. Nixon, but he did not engage in partisan political activity on so visible and sustained a scale again. But there were others to carry on, for one legacy of the revival of the fifties was a modernized conservative civil religion, reshaped from its Prohibition-era style by the addition of the metaphysical New Thought perspective, to take on the challenge of specific interest-group issues well into the Reagan era.

Notes

1. NVP to William Canfield, Sept. 24, 1956, NVP Ms. Coll.

2. Robert Wuthnow, *The Restructuring of American Religion* (Princeton, NJ: Princeton University Press, 1988), p. 322.

3. Interview with NVP, February 24, 1981, Pawling, New York. Peale quoted the Tawney-Weber thesis regarding religion and the rise of capitalism as evidence of Wesley's social success in promoting "the rise of the English working class." He was unfamiliar with E. P. Thompson's work other critical works in the field. In a column he wrote for *The American Way* entitled "Will America Go Leftist?" Peale made a lengthy reference to Wesley as the man who saved England from a French-style revolution. He argued, "This revival saved England from the violence of revolution. It

changed men so that they no longer were dissipated but hard working and frugal, saving their money and putting it back into business and so becoming property owners and employers. . . . In short, it opened the way to the greatest production of wealth in history." This column appeared in a conservative labor publication, *Labor News*, Feb. 23, 1945: NVP Ms. Coll.

4. *The New York Times*, March 28, 1934.

5. *The New York Times*, March 12, 1934.

6. Cited in Robert Moats Miller, *American Protestantism and Social Issues* (Chapel Hill: University of North Carolina Press, 1958), p. 122.

7. Ibid., p. 121.

8. May, M.A. *Education of American Ministers* (New York: Institute of Social and Religious Research, 1934).

9. New York *Sun*, December 22, 1942.

10. David H. Bennett, *Demagogues in the Depression* (New Brunswick, NJ: Rutgers University Press, 1969), pp. 5–8.

11. Norman Vincent Peale, "Organized Leadership: The Story of the Committee for Constitutional Government" (Committee for Constitutional Government, 1944). In this pamphlet, Peale made the connection between Gannett's original Borah support group and the Committee to Uphold Constitutional Government. NVP Ms. Coll.

12. *U.S. Congressional Record*, January 16, 1945, pp. 3–10. Rumely's past was routinely rehearsed in papers antagonistic to the committee, publications such as *The New York Post, PM*, and the un-self-consciously partisan *Friends of Democracy's Battle*. The *Congressional Record* revealed that neither pro- nor anti-New Deal members of Congress made an effort to conceal their partisan sentiments.

13. *U.S. Congressional Record*, May 15, 1944, Appendix, A2543, 2544. Gerard claimed Rumely handled the money, and Rumely said that Gannett was in charge of planning.

14. Special Committee Investigating Campaign Expenditures, typescript report, Oct. 25, 1940; NVP Ms. Coll., Box 183.

15. In 1944 when Peale's name was added to the advisory board of Spiritual Mobilization, the following were also members: economist Alfred Haake, politician Albert Hawkes, statistician Roger Babson, biblical scholar Edgar Goodspeed, Carlton College president Donald Cowling, University of California president Robert Gordon Sproul, and Stanford University president Ray Lyman Wilbur. There was also one woman on the board, Mary E. Wooley, Mt. Holyoke College president emerita.

16. James Fifield, "The Cross vs. the Sickle," sermon published by First Congregational Church, Los Angeles, August 3, 1947; NVP Ms. Coll. Fifield gave the sermon in other places, and the printed sermon itself seemingly enjoyed fairly wide circulation.

17. The *Defender*, Aug. 1944, p. 13; NVP Ms. Coll.

18. James Luther Adams, Liston Pope, Edwin McNeill Poteat, Guy Emery Shipler, Alfred Wilson Swan to NVP, Aug. 24, 1944; NVP Ms. Coll., Box 196.

19. NVP to Liston Pope, Sept. 5, 1944; NVP Ms. Coll., Box 196.

20. This is Wuthnow's argument in *Restructuring of American Religion*. See p. 114.

21. "Spiritual Mobilization," published by Mobilization for Christian Ideals, Inc., No. 147; NVP Ms. Coll.

22. NVP to James Fifield, Aug. 11, 1947; NVP Ms. Coll., Box 216. Peale's comment was: "If such a letter [criticizing the Council] as you propose is to go out I feel that it would put me in a very embarrassing position with regard to my broadcasting relationship. I feel that I would prefer to talk with Sam Cavert [executive secretary of the Council] privately before any letters are sent to anybody. I have known Sam for a long time, and I hold him in high respect, and I consider him a personal friend." Peale's counsel, given in "strictest confidence," was one of restraint. Also James Fifield to Donald J. Cowling, May 5, 1948; NVP Ms. Coll.

23. *The New York Post*, September 7, 1944. Pictures of Peale and Fifield appeared along with the column, with Peale identified as "veteran enemy of the New Deal. Peale is a leader of the many isolationist, anti-Roosevelt 'Constitutional Committees' financed by Frank Gannett, the conservative Republican publisher."

24. NVP to Hugh Thomson Kerr, Jan. 6, 1947; NVP Ms. Coll. It should be noted, however, that Peale concluded his letter by commenting, "Personally, I hold Dr. Fifield in very high regard as a sincere, earnest and capable leader."

25. NVP to Dr. James Claypool of the American Bible Society, July 15, 1947; NVP Ms. Coll.

26. In letter of Wesley Goodson Nicholson to Ralph Walker, July 22, 1948; NVP Ms. Coll.

27. James Fifield to Donald Cowling, May 5, 1948; James Fifield to NVP, April 26, 1948; NVP to James Fifield, January 13, 1949; NVP Ms. Coll. These communications make no mention of the name of the new organization, since it went unnamed until the organizational meeting in April 1950. The names of the planners, the objectives, the desire to include only the "like-minded" make it clear that the desired outcome is a relocated, somewhat transformed version of Spiritual Mobilization. Fifield encouraged the move, although his communications suggest a tinge of regret that Peale, with a patron like Pew, would not face some of the pioneering problems he encountered, and a sense, too, that perhaps his early trials were not fully appreciated.

28. Pew's extensive financial support for this project may have been what Fifield worried about when he wrote him suggesting that "unscrupulous leftist elements" might exploit Pew's role if they got wind of it. See Chapter V, fn. 63.

The evidence suggests that Pew paid the rent on office space for Spiritual Mobilization, at least in New York, Chicago, and Los Angeles. When it was time to renew the leases on the offices, Fifield wrote to Pew to ask if he expected "any change in the picture which would make it unwise for us to sign such leases through 1949? . . . I do not wish to bother you with details, but I need counsel and guidance." Fifield to Howard Pew, Oct. 6, 1948; NVP Ms. Coll. With probably similar considerations in mind, Fifield wrote to Peale saying that he hoped "Mr. Pew will send us five thousand dollars this month" and wondering whether Peale had any more definitive information about the prospect. Fifield to NVP, Dec. 12, 1949; NVP Ms. Coll.

29. NVP to Howard E. Kershner, President, Christian Freedom Foundation, Inc., Aug. 29, 1950; NVP Ms. Coll.

Both men were pleased, however, to have Pew's support. On December 13, 1948, Fifield wrote to Peale that it was wonderful that he had "won the confidence of our mutual friend, Mr. Pew," whose "assurance of backing" would free Peale from the kind of financial worries Fifield had had to cope with. Fifield to Peale, Dec. 13, 1948; NVP Ms. Coll.

The following year when Fifield wrote to Peale about the move of Spiritual Mobilization to New York in reconstituted form, he said he understood Peale's interest in making "an entirely fresh start—not using our name, Spiritual Mobilization, or in any other way seeming to be related to our movement." He applauded both the decision and the fact that Peale had secured the support of Pew as an "ally." Fifield to Peale, May 9, 1949; NVP Ms. Coll.

30. In a letter written with the hope of persuading Peale to work for Spiritual Mobilization, Fifield observed that a paid position with Spiritual Mobilization would allow him to give up his "productive lecture program" and could even be an "answer to a prayer." Fifield to NVP, Apr. 26, 1948; NVP Ms. Coll.

31. NVP to Christian Freedom Foundation, Howard E. Kershner, President, May 8, 1950; Kershner to NVP, May 9, 1950; "Report on the organizational meeting of the Christian Freedom Foundation," Hotel Wellington, New York City, April 17, 1950; NVP Ms. Coll., Box 239.

32. NVP to Howard E. Kershner, Oct. 9, 1962; NVP Ms. Coll., Box 154.

33. NVP to Howard E. Kershner, May 19, 1954; NVP Ms. Coll., Box 347. Peale confessed to being distressed by the "violent attacks" on the United Nations, and while admitting it had some weaknesses, asked rhetorically, "What are you proposing as a substitute that would give promise of ultimate peace?"

34. NVP to James W. Fifield, December 9, 1950. Fifield's response was that all he knew of *Christianity and Crisis* was that Reinhold Niebuhr was on its board, which seemed reason enough to him to not take either the magazine or Brown seriously. His view was that "After sixteen years of pressure and abuse from 'pinks' who are critical of what you and I believe . . . I have come to the conclusion that it is not worth the effort or time to try to recognize them." He concluded his letter by complimenting the Peales on their new apartment and with hopes that they would soon be able to engage their "chauffer [*sic*]." Fifield to NVP, Jan. 3, 1951; NVP Ms. Coll.

35. NVP to Joseph Edison, July 30, 1951; NVP Ms. Coll.

36. NVP to Stanley and Dorothy Kresge, Aug. 11, 1951; NVP to Dr. Henry P. Judd, Aug. 13, 1951; NVP Ms. Coll.

37. NVP to Joseph Edison, July 30, 1951; NVP Ms. Coll.

38. NVP to James Fifield, Aug. 2, 1951; NVP Ms. Coll.

39. NVP to the Rev. Richard Upshur Smith, Aug. 13, 1951; NVP Ms. Coll.

40. NVP, "Compulsion Is Un-American," Aug. 16, 1944. (The document is noted as being released by George Peck, head of a lecture bureau.) NVP Ms. Coll.

41. NVP to Morton R. Cross, Nov. 21, 1950; NVP Ms. Coll. Like many other letters in the NVP Manuscript Collection, this seems to be addressed to a person Peale did not know well yet whose interest he desired. Later on in the same letter he observed that the rank and file of labor was "heartily against" communism, and that not "every strike is inspired by Communism."

His problem of speaking for approval reasserted itself when a labor leader whom he knew through religious circles, John Ramsey, told Peale how impressed he had been with Walter Reuther's closing statement at a CIO convention: Peale's response to Ramsay was that "It is certainly a very wonderful Christian statement; high-minded in every particular, and is worthy of the eminent and outstanding leadership which Mr. Reuther has given over the years." NVP to John Ramsay, Dec. 22, 1955; NVP Ms. Coll.

42. Friends of Democracy, *The Propaganda Battlefront*, October 1944; *PM*, October 16, 1944; NVP Ms. Coll.

43. John Roy Carlson, *Under Cover* (New York: E. P. Dutton, 1943), p. 475.

44. NVP to E. B. Macrae, President, E. P. Dutton, July 1943; NVP Ms. Coll., Box 100.

45. NVP to John Roy Carlson, Sept. 27, 1943; NVP Ms. Coll., Box 194.

46. *The New York Times*, July 19, 1943.

47. John Roy Carlson, *The Plotters* (New York: E. P. Dutton, 1946), pp. 296–97.

48. George Seldes to NVP, Aug. 21, 1944; NVP Ms. Coll., Box 200.

49. NVP to Ernest Kornfeld, District Representative, South Jersey Industrial Union Council, Dec. 7, 1944; NVP Ms. Coll, Box 194.

50. Carlson, *The Plotters*, pp. 297–98.

51. NVP to "Dear Dad," Oct. 17, 1940; NVP Ms. Coll., Box 190. He wrote in a similar vein to a person who attended his church: "I believe it is in the best tradition of the American ministry to be actively interested in the cause of freedom and Constitutional government, which at this particular time is distinctly imperiled." NVP to F. Kraissl, Feb. 17, 1944; NVP Ms. Coll, Box 194.

52. NVP to Merwin K. Hart, Nov. 1, 1943; NVP Ms. Coll.

53. NVP to Channing Pollock, Mar. 17, 1943; NVP Ms. Coll.

54. The four books he suggested were Hartzell Spence, *One Foot in Heaven*; Richard Llewellyn, *How Green Was My Valley*; Lord Tweedsmuir, *The Pilgrim's Way*; and Alice Duer Miller, *The White Cliffs of Dover*, in the *New York Herald Tribune*, January 13, 1941.

On another occasion, when the Religious Book Publishers asked him to select twenty-five works from a list of a hundred books for Lenten reading, he had no problem suggesting titles and managed to include two by well-known liberals, Harry Emerson Fosdick and frequent Peale critic G. Bromley Oxnam. Oxnam wrote Peale to thank him for including his Yale lectures, *Preaching in a Revolutionary Age*. NVP, "Suggested List of Lenten Reading," 1945, typescript; NVP Ms. Coll., Box 202. Also G. Bromley Oxnam to NVP, Mar. 23, 1945; NVP Ms. Coll., Box 202.

55. NVP to E. F. Hutton, Dec. 10, 1943; NVP Ms. Coll.

56. NVP to Benjamin P. Vanderhoof, Nov. 3, 1943; NVP Ms. Coll.

57. Edward Rumely to NVP, June 14, 1941; NVP Ms. Coll., Box 183.

58. NVP to Edward Rumely, Mar. 30, 1944; NVP Ms. Coll., Box 195.

59. Rumely told contributors to the committee that he offered the Congressional investigators a list indicating 25,000 supporters, but that he had refused to single out the "the names of 224 who had contributed $100 or more." Edward A. Rumely, draft letter to Contributors to Committee for Constitutional Government, Feb. 7, 1945; NVP Ms. Coll., Box 201.

60. E. A. Rumely to "The Trustees" of the Committee for Constitutional Government, Apr. 19, 1946; and Edward A. Rumely to Neil Burkinshaw, Apr. 18, 1946; NVP Ms. Coll., Box 207. Rumely thanked Burkinshaw for his support as follows: "Indicted, you stood at my side unwavering; at every step of the way, through the first trial that ended in disagreement, through this trial to complete victory." He noted that only five hours earlier, after a twenty-five-minute deliberation, the jury had decided for acquittal. According to a *PM* article of October 15, 1946, the first trial had ended inconclusively because of the judge's ambiguous definition to jurors of the term "politics."

61. The committee volunteered this information to the Congressional investiga-

tors. Norman Vincent Peale, *Organized Leadership: The Story of the Committee for Constitutional Government*, published by the Committee for Constitutional Government, 1944; U.S. House of Representatives, *Congressional Record*, January 18, 1945, p. 10. (Copy in NVP Ms. Coll., Box 201.)

62. Friends of Democracy, *Friends of Democracy's Battle*, Feb. 1950, VIII, nos. 3 and 4; NVP Ms. Coll.

63. Edward Rumely to the "The Trustees" of the Committee for Constitutional Government, July 18, 1944; NVP Ms. Coll., Box 195.

64. *New York Journal-American*, November 13, 1951. Pegler alleged that Pennsylvania Congressman Frank Buchanan, recently deceased, had spearheaded the investigation of Rumely.

65. NVP to Dr. Edward A. Rumely, May 25, 1945; NVP Ms Coll., Box 201.

66. NVP to T. O. Thackrey, Editor, *The New York Post*, Apr. 16, 1946; NVP Ms. Coll., Box 206.

67. In a letter to a Methodist clergyman who quizzed him about his political affiliations, Peale wrote, "I do not consider myself a conservative, by any means." A month later he wrote to a Republican Congressman, "I have always been a most interested Republican and thoroughly believe in the principles of the party," but then he went on to advise the politician that the CIO knew how to talk to ordinary people, and unless the Republicans did the same, "the party is through." To engender more support for the free-enterprise system, Republicans needed to "know how to talk the language of the average man." NVP to Dr. Lyndon Phifer, Oct. 15, 1945; NVP Ms. Coll., Box 201. NVP to Honorable W. Sterling Cole, Nov. 23, 1945; NVP Ms. Coll.

68. American Committee for Spanish Freedom, Committee press release, Dec. 19, 1944; NVP Ms. Coll.

Peale repeatedly negotiated his own personal truce with the National Council of Churches (NCC). Ruth Peale was an active participant from the beginning. Norman Peale was essentially opposed to the concept of ecumenism as advanced by the NCC, its liberal policies frequently at odds with his own position on issues. Yet he defended the NCC from its worst critics. To a person concerned about the "socialistic" implications of the National Council, Peale wrote, "The last persons in the world who would ever be party to the adoption of a socialistic order are Mr. J. Howard Pew and Mrs. Peale. In fact, Mr. Pew has for many years been one of the most sturdy, thoroughgoing and aggressive defenders of freedom whom I have ever known in my life. I believe that it is a happy augury for the future when a man of Mr. Pew's caliber and sturdy faith, both religiously and politically, should be made chairman of the Laymen's Committee of the newly formed National Council of Churches." NVP to Dr. Arthur Blazey, Feb. 16, 1951; NVP Ms. Coll. The NCC also supported Peale's radio program.

69. H. L. Hunt, "Presenting Facts on Facts Forum," article 1, and "Presenting Facts on Facts Forum," article 2, *The Dallas Morning News*, May 11, 1953; May 12, 1953.

70. Lew Foster to H. L. Hunt (labeled "Copy"), Nov. 4, 1951; NVP Ms. Coll. The letter writer noted that the presidential election was exactly a year away and that in Peale's many contacts with audiences across the country he warned of the dangers the nation faced. He also noted the growth of *Guideposts* and thought that the magazine would be willing to run a story on Facts Forum.

71. Marble Collegiate Church office memo (initialed avm) for Dr. Peale, Dec. 7, 1951; NVP to H. L. Hunt, Dec. 10, 1951; NVP Ms. Coll.

72. H. L. Hunt to NVP, Dec. 14, 1951; NVP Ms. Coll. Hunt had figured out that 80 percent of the delegates to the forthcoming convention were "repeaters" from the last convention and that all delegates should be called on. He also had a plan for working with "Southern Constructives" to force a realignment of the two parties. If these southern Constructives, states righters, could not command use of the Democratic Party label, they would take another. H. L. Hunt to NVP, Mar. 13, 1952; NVP Ms. Coll.

73. George S. McMillan to Congressman Ralph W. Gwinn, June 23, 1952. The writer thought that sending the book to voters was a "cheap political trick" and a cover for a Taft candidacy. He believed the effort would backfire in Eisenhower's favor. NVP Ms. Coll.

A conservative, Gwinn was a friend of Peale's and a near-neighbor.

In 1958 Prentice-Hall, Peale's publisher, asked him to approach MacArthur with a request for his autobiography. In making the request Peale wrote, "I doubt if you have any more loyal admirer in the United States than the writer of this letter." NVP to Douglas MacArthur, 1958; NVP Ms. Coll., Box 137.

74. NVP to Frank E. Gannett, Apr. 1, 1952; NVP Ms. Coll.

75. NVP to H. L. Hunt, Dec. 31, 1951; NVP to H. L. Hunt, Apr. 26, 1951; NVP Ms. Coll.

76. H. L. Hunt, "Presenting Facts on Facts Forum," article 3, *The Dallas Morning News*, May 14, 1953.

77. NVP to Mrs. Hamilton (draft letter), Apr. 20, 1953; NVP Ms. Coll., Box 264. Also Olga Hamilton to NVP, May 20, 1953; H. L. Hunt to NVP, May 7, 1953. In Hunt's letter to Peale he said that in view of the critical letter from the Marble Church member, he hoped "that we are not getting you in too deeply." He admitted that in using the speaker about whom the Marble Church member complained that Facts Forum might have created some problems for itself. Hunt said he knew little about the speaker, although he had the impression that "communists generally hate him, but he may be quite vulnerable."

78. NVP to Dr. Louis M. Hacker, Dean, School of General Studies, Columbia University, Feb. 6, 1953; NVP to Ted Dealey, *The Dallas Morning News*, June 19, 1953; Dallas Gordon Rupe to NVP, Dec. 4, 1953; NVP to J. Wallace Hamilton, Mar. 19, 1954; NVP Ms. Coll., Box 347. His old friend Hamilton wrote to him: "Some of us are awfully disturbed down here, to see your name linked with this." He continued, "I was in Dallas, Texas, when this thing was hatched, and didn't like it then, but I like it a good deal less now."

79. NVP to H. L. Hunt, Nov. 16, 1953; NVP Ms. Coll.

CHAPTER 7

The Demise of Tribal Politics 1955–1985

> The boundaries of personal influence it is impossible to fix, as persons are organs of moral or supernatural force. Under the dominion of an idea, which possesses the minds of multitudes, as civil freedom, or the religious sentiment, the powers of persons are no longer subjects of calculation. A nation of men unanimously bent on freedom, or conquest, can easily confound the arithmetic of statists, and achieve extravagant actions, out of all proportion to their means.
>
> Emerson
> Essay on Politics
> 1844

Peale's withdrawal from Facts Forum was relatively painless, an unusual outcome given his desire to please. That he recognized a substantial risk to *Guideposts* by continuing with the controversial group made his decision easier. But it was also eased by the fact that he and H. L. Hunt had not been particularly close, and there was no real personal tie to break.

On the other hand, for Peale, politics, like religion, was a very personal matter. His fierce commitment to the Prohibition struggle sprang largely from a sense of tribal loyalty, although most Protestants generally supported the effort out of a similar, if more muted, awareness of group identity. With Peale, however, political attachments seemed more intimate even than tribal commitments, often more like family bonds. Politics was a kind of family affair, with candidates and the issues ideally viewed as public reflections of what were essentially familial concerns, about how to help the family to thrive and stay secure. It was a view of politics drawn from an earlier time, when Protestant hegemonic authority was taken for granted and cultural pluralism was a reality without power or voice.

Despite this almost primal pull of politics, however, Peale's renewed pledge in 1955 to stay out of the political arena forever had an authentic ring to it. He had acquired the status of a celebrity, and the demands of public life made it difficult to find time even for his family. He was hounded by newsmagazines and courted by television—profiled by Ralph Edwards on "This Is Your Life" and by Edward R. Murrow on "Person to Person"—as well as being portrayed in the movies, the subject of the Hollywood feature, "One Man's Way." And as the gains of the religious revival became more evident, he turned his attention to the expansion of his personal ministry. The relentless pressure of public life seemed to assure that he would keep his promise of political detachment, and at least for the time being experience political satisfaction vicariously through his recent ties to the National Association of Evangelicals (NAE).

The NAE agenda offered Peale the chance to recapture the kind of familial identity he associated with politics—and also religion. Most of the NAE leadership had transferred its Prohibition sentiments to the struggle over the Cold War, just as Peale had done, evidence that on matters global as well as domestic they were in harmony. When Graham and Peale symbolically joined hands at Madison Square Garden they were testifying to a relationship with important civil as well as religious implications.

His continuing conflict with critics over *The Power of Positive Thinking* facilitated Peale's openness toward the NAE. He brooded to friends about a "liberal party line" and the consequences visited on those who challenged it. Rebuked earlier by his adversaries for his politics, he now saw his theology up for attack. He became increasingly critical of "the organized church," particularly for its coolness toward the religious revival. He was decidedly ambivalent regarding the National Council of Churches, whose views were frequently at odds with his own but whose resources continued to support his radio program. It was reassuring that his new NAE colleagues welcomed him, if somewhat warily on theological grounds, enthusiastically on political ones.

Peale and the NAE understood their political kinship. Their common legacy from the prohibition struggle, a desire to restore Protestant hegemony, was grounded in an evangelical appeal to populist politics. These shared hopes, for a return to a mythic "early time" of Anglo-Protestant power carried on the strength of a populist insurgency, came together in the new evangelical civil religion taking shape. In the effort to give life to these hopes, however, Peale and his NAE colleagues knew that they could not count on the support of a united Protestantism. While the Protestant leadership and rank and file had been unified in the Prohibition struggle, that was no longer the case. The liberals and the evangelicals were divided theologically over social Christianity and a religion of personal experience, and they were separated politically over the way Cold War issues ramified in the culture.

These divisive tendencies reached a critical juncture in 1960, when the arguments of the 1928 campaign were revisited. The possible nomination of a Roman Catholic for President, and one perceived as a liberal, reconnected the political and religious issues once again, at a time when revival fires were just beginning to cool. The revival had helped identify a new coalition, represented by the two leaders of the movement—Peale and Graham—and including the leading figures within the NAE and its twin constituencies: a group of wealthy, largely self-made businessmen and the mass of "ordinary" people who subscribed to personal religion and conservative politics. In blocking the nomination of John F. Kennedy, the professed goal of the coalition leadership was to stop the erosion of religious freedom presumably attendant on the election of a Roman Catholic while at the same time striking a blow against a liberal Protestant elite that endorsed Democratic politics.

Politics: Personal and Local

Peale's participation in the effort to stop Kennedy's nomination proved to be his biggest blunder. He never fully recovered his pre-1960 stature, although some loss in eminence was probably inevitable as the revival began to wane. More seriously, while he had long since parted company with religious liberals, he had managed to stay closely identified with his own constituency. In opposing Kennedy on religious grounds, he alienated some of his own supporters and scattered some segments of the populist coalition.[1] For a populist leader, it was a critical misreading of his audience. The results of the campaign showed that even among conservatives and evangelicals, anti-Catholic bias was becoming, if it was not already, an unacceptable part of their conception of America's civil religion.

That Peale's earliest political lessons, about the value of personal and tribal loyalties, had an enduring influence was made clear in his interest-group affiliations, as with Fifield and Gannett, for example. But it was also revealed in his unbridled enthusiasm for particular candidates. This was especially evident in the case of Thomas E. Dewey and then Richard M. Nixon, though it could also be detected with Ronald Reagan, whose presidency occurred at a time when Peale was well advanced in years.

Peale's dogged support of a candidate might be expected to arise from a deep personal attachment to the political hopeful. That was usually but not always true. On some occasions he supported a contender because the person had been recommended by a friend, as when in 1952 he responded to requests from H. L. Hunt and Frank Gannett to persuade Douglas MacArthur to announce for the Republican nomination, though his own sympathies lay elsewhere. Peale's personal history revealed many occasions on which his naiveté and need for approval allowed him to be taken advantage of by acquaintances. His own real preference in the 1952 race surfaced when he became an active supporter of the Eisenhower-Nixon ticket in New York City, a campaign that allowed him to renew an old friendship with the next Vice President.

His politics of personal engagement, developed most fully in his relationship with Nixon, were first tested in the forties in Dewey's two campaigns. Peale had actively supported all the anti-New Deal candidates from Hoover on, so that his commitment to Dewey readily followed in that tradition. But as with Nixon, there was another, more personal dynamic involved. That Dewey was also a Quaker Hill neighbor in Pawling simply strengthened the connection.

By the time of Dewey's 1948 campaign, Peale was ready to engage in a full throttle effort for his neighbor, whom he had come to regard as a friend. In the spring of the year he wrote letters to members of the clergy in Oregon, where Dewey was entered in the primary, pointing out that Dewey was a "very sincere and earnest Christian," regularly in church in Pawling, and one who "could always be counted upon as President to know and to stand for that which is best in our Christian idealism."[2] He urged them to do what they could to promote Dewey's interests. That the dark horse Truman managed a narrow victory surprised Peale as much as it did the rest of the nation. The results, he told a friend, were a keen disappointment:

> I was at the Governor's headquarters election night until 4:30 in the morning. All of his friends from Pawling were there and what started out to be an evening of great triumph turned into a rather funereal experience before the night was over. We were all planning to go to Washington for the inauguration.[3]

Indeed, a columnist at the *New York Herald Tribune* had predicted in early November that a Dewey victory would bring to the White House a new social crowd from Pawling, and he named as examples Lowell Thomas; Mr. and Mrs. Raymond Thornburg of *Guideposts*; Fulton Oursler of *The Reader's Digest*; and the Peales.[4]

Peale was predictably loyal to his Quaker Hill neighbor. When Dewey decided to run again for governor of New York in 1950, he had to settle a score with his lieutenant governor, Joe Hanley, an aging ex-minister with a yen for gambling on real horse races, who had been promised a shot at the governorship himself. With little to recommend him for elective office other than an amiable personality, Hanley was asked by the Republican leadership to exchange a race for the governor's office for a more daunting fight for the U.S. Senate, a challenge he ultimately lost to Herbert Lehman. Early in the contest, however, to help out Dewey, Peale wrote on the stationery of Marble Collegiate Church a "Dear Fellow-Minister" letter to clergy in the state, pointing out Hanley's credentials as an ordained minister and observing that along with Dewey he had helped preserve "religious freedom" in New York.[5] The letter produced sharp rebukes from ministers who thought Peale's letter was out of line, particularly his implication that Dewey had done something special for religious freedom. Nevertheless, Peale remained close to Dewey the politician, boosting his programs, and even, as in 1956, appearing on his television show just prior to elections to criticize the Democratic campaign.

He also got involved in local races, where personal considerations were often primary. His endorsement of the candidacy of conservative Congressman Ralph Gwinn especially fit that criterion, since Gwinn represented neighboring Westchester County and was a member of Peale's own denomination, the Reformed Church in America. But he also assumed an identification with a candidate for judge of the court of special sessions in New York City, presumably because the man was an active layman in the city.[6] And a personal sense of outrage encouraged Peale to speak out from the pulpit against a New York City mayoral candidate because of his left-leaning labor views.[7]

In the second half of the decade of the fifties, as he and Billy Graham harvested the evangelical vineyard, they combined political values with expressions of faith. In 1960, these concerns coalesced around a program to prevent the nomination of the Roman Catholic Senator John F. Kennedy.

Shoring Up the Foundations

Events during the summer and fall of 1960 persuaded Peale to abandon his pledge to stay aloof from partisan activity, and he entered into the thick of the political fray. Evangelicals who opposed Kennedy's impending candidacy attempted to ground their challenge in terms of church-state principles, but few people were seriously misled by the strategy. Despite Will Herberg's description of the nation as a melting pot of Catholic-Protestant-Jewish beliefs, interaction across religious boundaries was infrequent and often tense. In the years before Vatican II, Protestants and Catholics in particular shared a mistrust of each other: Catholics held fresh memories of bigotry and intolerance, and Protestants generalized about the insular, authoritarian nature of Roman orthodoxy. For Protestants of Peale's generation, there was also the notable legacy of 1928 and Al Smith's campaign. One did not need to look far to find evidence of suspicion existing between fellow Christians; in that regard, Vatican II and Kennedy's election were historic breakthroughs.

Peale had hoped to create the impression that his was an ecumenical ministry. Working in the pluralistic environment of Manhattan, he had participated in public displays of ecumenism, although he disliked the politics and bureaucracy of the organized ecumenical movement. He cooperated in interreligious undertakings and hoped that *Guideposts* would be accepted as the interfaith magazine he intended it to be. But old values and stereotypes were not readily ripped out root and branch. Dormant embers of anti-Catholicism might be fanned again, as they were by the prospect of Kennedy's candidacy, and the damage extended to his ecumenical projects.

In a 1955 interview with *The New York Post* about his politics, Peale had said this: his political preferences had been "somewhat left-wingish"; his past political involvements had become a cause of regret; and his previous forays had been the result of his "amiability." He explained that he "always had a tendency to do something which friends asked me to do, . . . But no more. Not even my best friend in the world could get me to join."[8] Certainly five years later, he wished he had held to that Shermanesque pledge. Of course, *The Post* readily documented the disparity between Peale's "left-wingish" claim and his membership in the National Republican Club, the old Committee for Constitutional Government, and his many pronouncements against labor and the New Deal. It also took notice of his denial of having written the Joe Hanley letter to "Dear Fellow-Minister" and the fact that a copy of the letter existed in media files. But the paper took seriously the preacher's disclaimer of future political engagement, observing that his "following is so broad that political disputation can only serve a negative purpose."[9] Unfortunately for Peale, the specter of a Catholic President became a kind of call to arms, and he abandoned his declared resolve to stay out of politics. He once voiced the wish to "withdraw to some cloistered place and let the world go by," but even as he spoke he seemed to realize it was an idyll desire.[10]

Can a Catholic Become President?

As the Democratic primary heated up in the spring of 1960 and candidates John F. Kennedy and Hubert H. Humphrey focused on the crucial states of West Virginia and Wisconsin, Peale made a speech in Charleston, West Virginia, a month before the vote. Arguing that it was not only relevant to the campaign but essential to it to raise the religious issue, he said the basic question was whether, if Kennedy were President, he would be as "free as any other American to give 'his first loyalty to the United States'"[11]

Interviewed by radio and television in Charleston, Peale brought up some Protestant lore that would continue to circulate throughout the campaign. According to the story, in 1950 as a congressman from Massachusetts, Kennedy had been invited to attend a fund-raising dinner in Philadelphia on behalf of the Chapel of the Four Chaplains honoring the clergymen of three different faiths who gave their life jackets to others during a World War II sea disaster. Kennedy initially accepted the invitation and then hastily withdrew. Peale, who had a special interest in the event because one of the chaplains was the son of his old friend and former Marble Church minister the Reverend Dr. Daniel Poling, reiterated the standard Protestant version, that Kennedy's withdrawal had been determined by the decision of Cardinal Dennis Dougherty. "If Kennedy could be told once [what to do], what would he do under other circumstances?" Dr. Peale asked. "Should any ecclesiastical authority be able to interfere with the freedom of a public official of the United States?" Kennedy's office responded with the alternate version, which claimed that Kennedy learned he had been invited not as a member of Congress but as a representative of the Catholic faith, and "as such, he had no credentials to appear at the dinner."[12] These differing accounts would be retold throughout the campaign.

As he typically did when speaking in cities around the country, Peale found the opportunity while in Charleston to have dinner with old friends. Writing later to thank them for their hospitality, he made a point to lecture his hostess, a Kennedy supporter, about her priorities. He urged her to cast her party vote for Humphrey. He said, "I don't care a bit who of the candidates is chosen except that he be an American who takes orders from no one but the American people." He then asked how a "dedicated Protestant as yourself could so enthusiastically favor an Irish Catholic for President of our country which was founded by Calvinistic Christians?"[13]

Many Protestant clergy were dismayed by what their colleagues were making of "the religious question" and sought to put it in perspective, if they could not put it to rest. As the polemics intensified, Frank Sayre, Dean of the Episcopal Cathedral in Washington, circulated a letter to Protestant ministers asking them to advise their congregations to exercise "charitable moderation and reasoned balance of judgment" when confronted with the issue.[14] In responding, Peale said it would be hypocritical of him to sign Sayre's open letter since he believed that airing the religious question would be therapeutic. He had "gotten very tired of the power hungry and contemptuous attitude of the Roman Catholic Church," he observed, and to sign Sayre's statement "would give the impression to thousands of Protestants that I am for Kennedy or at least openminded toward him, and that would be a falsehood." "I am sorry, Frank," he continued, "but I think it time to restrengthen the oldtime, if somewhat narrow, loyalties of Protestantism, or else we shall deteriorate altogether."[15]

Peale was correct in assuming that the good will of liberals, such as Dean Sayre, was not sufficient to still the passions generated by the religious issue. Indeed, given the level of tension in the campaign, it was predictable that even seemingly trivial matters would arouse partisan ire. Daniel Poling, an outspoken moderate evangelical and an experienced politician, had regular access to a diversified audience because of editing the Protestant monthly *The Christian Herald* and also because he preached regularly for Peale at Marble Church's Sunday evening service. When the Catholic weekly journal *America* ran an article entitled "Post Protestant Pluralism," Poling objected to the designation and language that implied that the country was "Post Protestant." He urged people to ignore religious affiliations when voting for candidates, but to be certain to avoid voting for the individual who "would accept the dictation of his church or faith" when executing the duties of his office.[16] Peale, too, was distressed by the phrase "Post Protestant," but even more concerned when Robert F. Kennedy, referring to his brother's pluralities in West Virginia and Wisconsin, spoke of the fairmindedness of "non-Catholics." In a letter sent simultaneously to the Kennedys and the newspapers, Peale suggested that the term "non-Catholics" was deprecatory, with "the implication of superiority," and therefore insulting to "the majority of people in this country."[17]

Fully engaged in the conflict by May, Peale collected his feelings about the election in a sermon at Marble Church. Designed as an evangelical critique of the social gospel, the message attacked the "highly intellectualized" "political mechanism" that had become modern Protestantism. Organized Protestantism, he charged, had become "undemocratic," with small groups issuing statements that claim to speak for all Protestants. It had become a blend of "humanitarianism, socialism and every other nicey-nicey-ism," so broad that it had become shallow. "The only true Protestants left in the United States are those who believe in the Bible and Jesus Christ the Savior and in salvation from sin," he asserted. What was needed in the current hour, he said, was "the old, strong, narrow Protestantism that made the United States strong."[18] The sermon, with its strike at the liberal establishment, was essentially a personal statement about his having made common cause with the NAE, and his new associates received it as such.

During the summer of 1960, as the campaign frenzy continued, Ruth and Norman Peale continued their typical pattern of spending the period from June to September abroad, which usually involved traveling in Europe. Through his office and newspapers, Peale kept in touch with developments in the campaign. They were decisive months in the struggle for the presidency, and Norman was no less keenly concerned with the outcome for being thousands of miles away.

Among notable evangelicals, like Harold J. Ockenga of Park Street Church, Boston, and L. Nelson Bell, Billy Graham's father-in-law, the election was so vital a matter that they had their sermons on the topic printed and circulated.[19] There was a flood of material of the same genre. The matters these sermons addressed seemed more temporal than theological, focused on questions of power and politics within what was described as a monolithic, hierarchical institution. In July, while visiting Austria, Peale outlined his worries in a letter to his friend, Dan Poling, who responded by reflecting Peale's concerns, though sounding much less anxious. Poling was, he said, satisfied with Kennedy's answer to his question about whether he would be controlled or not by ecclesiastical policy, and he recommended that any organized Kennedy opposition within religious circles await Peale's return in September. He also commented, "I have not seen Billy Graham since we were together in Washington, D.C. when we heard him say what he did about taking the stump for Nixon." He thought that a timely public statement signed by Peale, Graham, the Reverend Samuel Shoemaker, and himself would be more effective than a "broadside with many signatures."[20]

Still on holiday, Peale's distress increased. His wife shared fully his uneasiness and considered taking to the stump to support Nixon, saying she was worried about Kennedy's "complete incompetence, lack of judgment and emotional instability." Although her husband's position prevented his doing this, she believed that "as a wife, mother, and citizen" she had that option.[21]

Instead, she and Norman chose what they thought was a less confrontational approach. Ruth drafted a letter to a friend, a printer in Connecticut, to learn the cost of printing 350,000 copies of an excerpt from a sermon her husband intended to preach in Marble Church on September 18 shortly after their return to New York. No ad-libbed address, this was to be a special sermon, entitled "What a Protestant Should Do Today." The printed statement, paid for by the Peales themselves, would be distributed to the press, to the FCL mailing list, and to the Marble congregation on September 18. "To say that we are concerned about the coming election is putting it very mildly," she wrote.[22]

Also in her letter to the printer she referred, almost parenthetically, to a meeting Peale had had the previous day, a meeting with great consequences not only for her husband's future but for the future of the campaign. "Norman had a conference yesterday at Montreux, Switzerland," she wrote, "with Billy Graham and about 25 church leaders from the United States. They were unanimous in feeling that the Protestants in America must be aroused in some way, or the solid block Catholic voting, plus money, will take this election."[23] She noted that they would arrive in New York on September 6. Although the prospects were discouraging, they clearly believed that an aroused Protestant electorate might yet save the day for Nixon.

Peale surely expected his proposed sermon on "What a Protestant Should Do Today" would motivate its readers. The statement may never have been circulated: The sermon was destined not to be preached. More a thinly disguised piece of campaign literature than a sermon, the statement attempted to distinguish Protestantism from Catholicism, defining Protestantism as "a thoroughly democratic faith" that repudiated "an authoritarian hierarchy" and a "supposedly infallible man." What followed were ten provocative suggestions to Protestants: about remembering the martyrs who kept the faith, at the stake, in the fire, "chained at low water mark, perished in the rising tide, but who kept the faith, singing hymns as they died"; about refusing to be called a bigot because "you are firmly attached to the Protestant faith"; about realizing that three out of four Americans are Protestant and therefore should act in a unified way; and about recalling that "American freedom grew directly out of the Protestant emphasis upon every man as a child of God," and that "Protestantism and Freedom were married in Geneva, and John Calvin performed the marriage ceremony."[24]

A Friendship with Nixon

That Peale was an old friend of Richard Nixon magnified his interest in the outcome of the race. Peale and the Vice President had a long history together, dating back to the early 1940s when Nixon was in the Navy and stationed in the New York area. Although he regularly identified himself as a Quaker, Nixon worshipped at Marble Collegiate Church during the war, and the Peales and Nixons became acquainted. The friendship lapsed, but then resumed in the fifties when Nixon moved to Washington following the Eisenhower victory. In fact, Peale may have privately taken some small credit for the move, for despite his early meeting with MacArthur, he had become a leader of the Eisenhower-Nixon bandwagon in New York once their nominations were assured.[25] Over the next eight years, Peale and Nixon stayed in contact, with the Nixons attending Marble Church whenever they went to church in New York City. And Peale would make it a point to write a glowing letter of support if he heard a Nixon speech or statement that impressed him. And of course he sent along a copy of *The Power of Positive Thinking*, which Nixon pointed out had been recommended to him by Mrs. Earl Warren.[26] The relationship was sufficiently familiar to members of the business-civic club network in which Peale traveled that he often had requests from program chairmen to use his good offices with Nixon to secure him as a speaker.

By the time the election year 1960 opened, Peale was a firm Nixon partisan. In March, speaking as a guest at the First Methodist Church in Los Angeles, Peale had Nixon's mother in the audience and then sat with her after at a luncheon. The following day Hannah Nixon wrote to him to express her appreciation of the event saying, "I understand why Richard is such an admirer of you."[27] That occasion unfolded as an opportunity for an exchange of letters between the two men, allowing Peale to remind the Vice President of "how deeply interested I am in your present campaign" and of his "earnest prayers . . . that victory will crown your efforts."[28] Nixon answered with a gracious note, mentioning that Peale could be expecting as a gift from him a copy of the highly flattering account of *The Real Nixon* by Bela Kornitzer.[29]

The book arrived along with the heat of the summer campaign. Nixon invited Peale to give the prayer at the session of the Republican Convention that nominated him, but Peale declined, noting his European sojourn, including a longstanding arrangement to be at the Oberammergau Passion Play on that date.[30] He did, however, volunteer to send copies of the Kornitzer book to his "opinion-forming" influential friends, along with a cover letter that attempted to sound neutral. The letter concluded with the statement, "I do not enter into politics but it is my business in life to do what little I can to help people know and understand each other as brethren."[31] Nixon was understandably grateful for the help and hopeful that Peale hadn't received "too many brickbats for the very generous words you spoke in my behalf."[32]

In early August, from his vacation location in Europe, Peale sent candidate Nixon his thoughts on the campaign. He offered to do whatever he could to help the effort, sounding a note of foreboding when he commented, "as for possible brick-bats—we shall not worry about that. (So far none, however.)" He then went on at length about his concerns over the campaign and the Kennedy candidacy and mentioned that "Recently I spent an hour with Billy Graham, who feels as I do, that we must do all within our power to help you." And he also sounded his familiar populist note when he advised him to conduct a "Walking, man-to-man campaign down the streets of America" so that the "average, everyday, simple people . . . Bill, John, Mary, and Bertha" would get to know the real Nixon and not be taken in by the false Kennedy "glamour." Clearly worried about negative comments in the press about Nixon's personality, Peale noted, "Issues are important, but the contact of personality is more so. And your personality never fails to communicate when people get close to you."[33]

"The Peale Group"

By that point in the summer, the NAE had become organized and active in its campaign strategy. Donald Gill, a young Baptist minister, was given leave from his job with the NAE to organize a campaign interest group, which took the name of Citizens for Religious Freedom. Rather coincidentially, while these activities were occurring at home, the two leaders of the revival of the fifties, Peale and Graham, were once again summering in Europe. Graham had made plans for convening an evangelical summit in Montreux, ostensibly to deal with issues of general concern and not the campaign per se, and Peale eventually was included, either on his own or on Graham's invitation.

According to Ruth Peale's note, the Montreux meeting occurred on August 18, with about twenty-five of the leading evangelicals in attendance, including Harold Ockenga and Nelson Bell, in addition to Peale and Graham. Just what conclusions were reached at the meeting became a subject of serious debate among those involved, in Montreux and at home, for a long time after. The group met unobtrusively, almost secretly, never publicly mentioning the meeting again. Because Peale bore the brunt of the negative public outcry that followed the group's exposure, his friends felt that Graham had abandoned him and them in a dilemma that Graham had actually created.

Given the charged atmosphere and the pace of events, there may have been slips in communication. The Montreux group apparently delegated Peale to arrange a meeting with Nixon and a select group of clergy leaders, namely Ockenga, Sam Shoemaker, Dan Poling, and himself.[34] In addition, the planners saw the need for a larger, public meeting, probably in connection with Gill's group, which would also take place in Washington (as presumably would the session with Nixon).[35] Dates for these gatherings would obviously have to await confirmation, but considering Peale's regularly frenetic schedule, it is likely that he and the others delegated to meet with Nixon attempted to schedule the two sessions very close together. Within a week, according to Daniel Poling—by August 25—the date of Wednesday, September 7 had been nailed down for the public meeting.[36] The Peales were to arrive in New York from Europe on September 6. Norman's plan may have been to preside at the public meeting on the 7th and hope that the session with Nixon could be arranged for the 8th or shortly thereafter.

Communication was sufficiently clear between the Montreux planners and the leadership of the NAE's Citizens for Religious Freedom, however, to allow Donald Gill, as the group's titular head, to develop the agenda for the public meeting rather quickly. By August 29 the list of participants had been drawn up, notwithstanding the fact that a number of the speakers had just been at Graham's gathering overseas. Gill was "delighted" that Peale was going to preside at the all-day session scheduled to meet at Washington's Mayflower Hotel. The other designated speakers included Daniel Poling, who was to give the keynote address; Ockenga; Nelson Bell; the Reverend Fred Nader of the First Baptist Church of Flushing, New York; and Dr. Glenn Archer, head of Protestants and Other Americans United (POAU), an activist lobbying group. Gill told Peale he expected representatives from the "National Council of Churches, National Association of Evangelicals, Southern Baptists, and other groups not related to any of these." "[O]ur plan," he said, is "to be fair, factual, and candid in expressing Protestant concern."[37]

When Peale arrived in Washington on the 7th, the meeting was indeed arranged as Gill had described it. There were printed programs, invitations, and packets of material designed to assist those attending with their thinking about the election. The program participants, in addition to Peale, Ockenga, Bell, Nader, Gill, and Archer, also included William R. Smith, a Church of Christ layman from Lake Village, Arkansas; Dr. Clyde Taylor, an officer with the NAE; and the Reverend Roy Laurin, from the Eagle Rock Baptist Church in Los Angeles. The token outsider was Dr. George M. Docherty of the New York Avenue Presbyterian Church in Washington, who gave the invocation. Docherty promptly dissociated himself from the group the following day, saying it was his intention to vote Democratic.[38] There were reportedly about 150 people, clergy and lay, in attendance. Unknown to the delegates at the time, two reporters were present.[39]

At the conclusion of the conference, it was Peale who met with the press. Helped out by Gill and Ockenga, he fielded questions from the Washington press corps, vastly different from his typical audience. He also made available copies of the group's statement, which observers said later had been prepared in advance of the conference. The statement was a five-point indictment of the politics of the Roman Catholic Church, the topic that had, in fact, served as the focus of the conference. It charged the Church with being a political as well as religious organization, which had been known to deny freedom of "conscience" in countries where it was "predominant." In the United States, the statement contended, the Roman Catholic Church had "attempted to break down the wall of separation of church and state," and cited the instance of nuns in clerical habits teaching in public schools in Ohio as an example. It questioned whether a Roman Catholic President would be able to withstand "the determined efforts of the hierarchy of his Church," and observed that the prohibition against his attending interfaith meetings would surely be a blow to "religious liberty." It concluded with the allegation, "Finally, that there is a 'religious issue' in the present political campaign is not the fault of any candidate. It is created by the *nature* of the Roman Catholic Church which is, in a very real sense, both a church and also a temporal state."[40]

Peale's casual, ad-libbing style did not stand him in good stead on this occasion of the Washington press conference. After characterizing the recently concluded session as a "philosophical" discussion of "the nature and character of the Roman Catholic Church," he had to admit under questioning that no Catholics had been invited to the meeting to explain their stand. Neither had Jews been invited, nor the two leading theologian activists in New York, Reinhold Niebuhr and John Bennett. Had Niebuhr been invited, said Peale, nothing would have gotten done. He accurately described those who did attend as members of "evangelical and conservative" groups. Asked whether he thought Nixon's Quakerism might pose problems for the Vice President, Peale answered that he didn't know that Nixon ever let his Quakerism bother him.[41] Peale said that he was not satisfied with Kennedy's promise to remain free of church influence in the exercise of his office, since he did not believe that Kennedy could distance himself in that way.

When asked who organized the meeting, Donald Gill responded that he had, with the help of a layman, J. Elwin Wright of Rumney Depot, New Hampshire. The press subsequently pointed out that Wright was an ordained clergyman, who was past co-secretary of the Commission for World Evangelical Fellowship and was a member of the NAE.

Peale discovered quickly how grievously he had misread the temper of the times. The following day, Drs. Niebuhr and Bennett, serving as vice chairmen of the Liberal Party in New York, issued a joint statement indicting "Dr. Peale and his associates" for "blind prejudice."[42] The conference immediately took on the designation of "the Peale group," and Peale was besieged by critics and members of the media. He and Ruth slipped out of town, and he remained in hiding for the next ten days while the critics gained the headlines.

The public response was overwhelmingly critical of "the Peale group." Roman Catholic Church leaders and theologians were very measured in their response, while Jewish groups were highly critical of what appeared as a "religious test" for public office. The most vociferous criticisms, however, came from liberal Protestant Church leaders, who, philosophically opposed to the "Peale group," were also embarrassed by its statements.

Within a week, Peale severed his ties to the Citizens Committee for Religious Freedom, explaining that he attended the conference only "as an invited guest," and that he had taken no part in the preparation of its statements.[43] On September 15 he issued a public statement, which he sent to his congregation and to the press.[44] The statement pointed to his naiveté for not realizing "that under the circumstances this would immediately get involved in the political campaign." Because of his involvement, "which actually is contrary to my nature," he said, he had isolated himself from the press, thereby making matters worse, since the media continued to refer to the conference as "the Peale group." His distress over this "unhappy situation" prompted him to offer his resignation to Marble Collegiate Church.[45] The board of elders and deacons refused to consider it.

The repercussions of the meeting were far more extensive than the summary dismissal of Peale's resignation by the Collegiate board would indicate. Radio and television broadcasters called for his appearance, to participate in panels and discussions with persons who wanted to question him on his views. He remained closeted as Ruth Peale and his secretaries dealt with his public. *The Philadelphia Inquirer*, the paper published by his personal friend Walter Annenberg, immediately dropped his syndicated column, the medium for his message most affected by the crisis. Over the next few weeks, several other papers followed suit, in a move that at the time seemed to him as a mass abandoning of ship, although apparently fewer than 10 percent of the 196 papers that carried his "Confident Living" column canceled him. He canceled all his scheduled speaking engagements himself, in some cases learning he had been dropped before he had the chance to withdraw. He terminated discussions on a planned movie, and he placed all other plans, except for publishing, in abeyance. He did not re-emerge to face his public until the following Sunday, September 18, the day on which he had originally planned to deliver his sermon on "What a Protestant Should Do Today." The sermon he delivered in its place was "What Christ Does for People," essentially an oblique admission he had made a mistake. The direction of the sermon had been suggested by the editorial staff of *Guideposts*, which had prided itself on its ecumenical tone, and saw the sermon as an opportunity to mend fences and to keep subscribers from dropping. The congregation at Marble Church was supportive, offering the unique tribute of a standing ovation as he entered the sanctuary at each of the two morning services on the 18th, Peale's first Sunday back.[46] He continued to avoid the press after the worship services while his wife spoke briefly with reporters. Nothing was said about the Montreux gathering, at the time or later.

A "Protestant Underworld"

John Bennett, dean of Union Seminary's faculty at the time, alluded to an unspoken agenda when he referred to the Washington conference as one example of a "Protestant underworld" at work. Although Bennett did not charge Peale personally with being the leader of this "Protestant underworld," the press quickly took up the designation, and it found expression in many quarters. Writing in the journal *Christianity and Crisis*, Bennett charged that a "religious opposition to Senator Kennedy of the type associated with a kind of Protestant underworld—an opposition that expresses itself in unsigned manifestoes and stirs up undisguised hatred of Catholics—is still with us."[47] In addition, he wrote, there was a new kind of attack on Senator Kennedy, one that pretended to be fair, to renounce bigotry, but that launched one-sided attacks on the "worst in Spanish or Latin American Catholicism," and put the issue of religious freedom in the worst possible light. Clearly, this latter charge was aimed at the so-called Peale group. Specifically, said Bennett, the attacks by "reputable Protestants, especially the statement of the group under the leadership of Norman Vincent Peale, show no understanding of the inner dynamics of the Roman Catholic Church."[48] Bennett allowed that there were Roman Catholic hierarchical "pressures" that he didn't like any more than the Peale group did, but he believed that a Catholic who "knew his way around in the Church" might be better able to handle them than a Protestant, who exaggerated the local authority of the Church. This was particularly true, he thought, in foreign policy, where the American Church, unlike the Vatican, was rigorously engaged in a conservative battle of the Cold War.

"What kind of country do these Protestants want?" Bennett asked rhetorically, "a country in which 40 million citizens feel that they are outsiders?" He found it curious, he said, that those who take the leadership in "the Protestant attack on the Roman Church" would not vote for a liberal Democrat whatever his religion. He added that the strongest opposition on the religious issue was centered in that part of the country which also opposed Kennedy on civil rights and economic policy.[49]

Peale remained silent, possibly too wounded to take on Bennett's challenge, but Dan Poling was a willing advocate for his friend. Speaking at the Sunday evening service on that same day, Poling observed that as vice presidents of the Liberal Party, Bennett and Niebuhr were well within their rights to support the presidential candidate of their choice. But they were out of bounds, he said, for calling other Protestant clergymen bigots and leaders of a Protestant underground, although he mentioned that a personal note from Bennett excluded Peale and himself from that particularly onerous charge. Poling pointed up the conflicting values in the dispute by noting Niebuhr's public comment that "Dr. Peale and I disagree on everything, religiously and politically."[50]

At the time, Niebuhr and Bennett represented the incarnation of the liberal Protestant theological establishment, especially to Peale. Union Theological Seminary in New York was the vital center for liberal activity, political and theological, and Niebuhr and Bennett, as exemplified by their roles in the Liberal Party, were the catalysts at its core. It would have been difficult to find two better representatives of cultural styles in conflict than Bennett and Peale. Ironically, Peale's only son, John, was at that very time a student at Union, where he sought Bennett's counsel while simultaneously serving as supporter and defender of his father.[51]

Liberal opinion in church and state lined up behind Bennett and Niebuhr in condemning the Washington meeting for introducing the religious issue. Robert Kennedy spoke for his brother, noting that both Peale and Poling were Republicans with ties to Nixon, which therefore cast suspicion on the sincerity of their concern for church-state issues. Governor Michael DiSalle of Ohio, responding to the charge of nuns teaching in public schools, asked for an apology or a retraction of the allegations; Senator Henry Jackson called on Nixon to repudiate the group. Former President Harry Truman said he was "disappointed" in Peale, and labor organizations added their voice to the Peale dissenters. Nixon himself said he believed that all talk of the religious question should be dropped, admitting he was unsure of just how Kennedy could do this. In religious circles, the Methodist *Outlook* magazine editorialized against the objectives of the Peale group, and *The Presbyterian Outlook* offered similar sentiments. Representative Adam Clayton Powell charged the Washington group with "reviving the spirit of the Ku Klux Klan," and *The Christian Century* said the conveners were "myopic" and guilty of "political irresponsibility." A published statement, signed by a hundred Protestant, Jewish, and Roman Catholic leaders and condemning the Peale group, included among its signers some longstanding Peale antagonists, such as Niebuhr, G. Bromley Oxnam, and Methodist Lloyd Wicke. Peale was reliving the fight over *The Power of Positive Thinking*, only now in politics.

Peale's rapid withdrawal from the group, on September 10, just three days after the meeting, brought him in for fresh criticism. Now those who had previously taken pleasure in the public challenge to Kennedy and the Democrats, which the conference posed, found additional cause to criticize Peale, this time for cowardice, for refusing to stand by the position he seemed to represent. In telephone interviews and in his sermon, he explained that he had acted like a "willy-nilly" man, but that "I was not duped—I was just stupid."[52] He went to the meeting, he said, innocently "like a babe in the woods," his real reason for going to Washington being to attend a "spiritual retreat" for "government leaders" conducted by Dr. Abraham Vereide on that same day.[53]

Within a week of the Washington meeting, Kennedy faced the religious issue head-on, giving a talk to an obviously unsympathetic Greater Houston

Ministerial Association and then fielding questions from the audience. He gained high marks from his listeners and the press for his forthright answers. When asked about Kennedy's performance, even Daniel Poling had to admit it was "magnificent."[54] Publicly at least, the bottom just dropped out of what had until recently seemed like a major evangelical challenge to the Kennedy candidacy. A whispering campaign continued, but Kennedy's Houston performance made public debate of the religious question unacceptable. Additional radio stations canceled Peale's program, and he considered resigning as editor-in-chief of *Guideposts.* Donald Gill and his former associates on the Committee of Citizens for Religious Freedom issued a statement absolving Peale from the role of architect or planner of the meeting and expressing concern that he was wrongly being singled out for criticism.

But the critical comments and harsh mail continued. He found little quarter in the press, surprisingly not even with some of the papers that were sympathetic or that published his column. The editor of *The Saturday Evening Post*, who identified himself as a Nixon man, said that he was "deeply disturbed" by the actions of the ministers at the Washington conference.[55] *The Pittsburgh Press* editor indicated the need to cancel Peale's column because of "the great number of letters and telephone calls received after the Washington meeting."[56] A less favorably disposed publisher of *The Detroit Free Press*, commenting after the election, allowed that while he did not think that Peale was "responsible for Nixon's defeat," he was convinced that the "so-called Peale episode was a factor in arousing the Catholics, and evidently a number of Protestants."[57] *The New Republic* reported that

> Highly reliable sources have linked the sponsorship of the conference with such prominent Republican personalities as J. Howard Pew of the Philadelphia Sun Oil Company Pews, who backs *Christianity Today* [edited by L. Nelson Bell, Billy Graham's father-in-law and one of the conference speakers] and who in the words of Reinhold Niebuhr "combines piety with right-wing Republicanism in a most remarkable way."[58]

No one else took up this specific charge, although the unfavorable commentary on Peale, with veiled imputations of conspiracy, continued. Peale became more distraught.

To his friend James Fifield he wrote, "I ruined myself. That's about it."[59] He tendered his resignation to the New York Rotary Club, continued to decline speaking invitations, and worried about Marble Church's ability to carry out its planned fund-raising campaign. To the working editor of *Guideposts* he wrote, "I have been hit very hard and I feel that I have been practically destroyed. . . . The church is my basic job, and I did not press my resignation, though who knows but what I may have to resign ultimately anyway."[60] Peale contemplated making a public statement urging support for Kennedy. He also indicated he was reconsidering his offer to resign as editor-in-chief of *Guideposts.* Ruth Peale and his close associates worried about his mental state, going so far as to shield him from critical mail, which Ruth felt "would throw him into another period of despondency without accomplishing anything."[61] To such an extremely sensitive person as Peale, all the unfavorable comments were hurtful, but those that were surely most upsetting were those that came from old Republican friends, especially when they suggested that the conference may actually have hurt Nixon's chances.[62]

By withdrawing from his customary whirlwind round of activity, Peale had more time to brood, and his depression increased. He became convinced himself that his actions created a sympathy vote for Kennedy, and weeks before the election predicted "without any doubt that Kennedy will be elected."[63] To an old friend from *Guideposts* he spoke of "irreparably impaired influence," describing himself as feeling "like a man who had a nice house which was hit by a hurricane, and all he was able to do was to extricate a few sticks and boards therefrom and build up a little lean-to in which to take shelter for bare existence."[64] What withdrawal from the public arena did provide him was an opportunity for uninterrupted time to write. And, compelled by a strong desire to get his feelings on paper, he determined to embark on a new manuscript, one that he warned his attorney was going to be his "final book."[65] Even long after the storm of unfavorable publicity had stilled, however, Peale continued to worry, fearful that he was "becoming emotionally unbalanced, [because] I cannot seemingly surmount in my own mind the problem imposed upon me by that furor which developed last September."[66]

To his trusted friend and partner at the Institutes for Religion and Health, the psychiatrist Smiley Blanton, he poured out his feelings. Reiterating his belief that "irreparable damage was done," he allowed a unique display of anger to surface. His post-Kennedy book, he said, was a statement about his irrevocable withdrawal from the realm of public political commentary, specifically, that "never again would I have anything whatsoever to do with the religious freedom issue—that I am through with it."[67] In addition, he had "come to the conclusion that all politicians, and I mean all, are phonies, and most of them are crooks." Although later he would obviously change his mind, he told his old friend that as far as politicians go, "I wouldn't be caught dead with one of them, no matter who he is." And his feelings about most ministers—meaning liberals—were similar: "I have been making quite a few attacks on the hierarchy of the Protestant Church, for I think they have killed the Protestant Church, or at least have dealt it a mortal blow."[68] Blanton tried to be reassuring. He disagreed with Peale's estimate about the extent of the damage, conceding, however, that as long as he continued to think that way, it would remain a disruptive factor in his life.[69] Sounding professionally therapeutic, he urged Peale to understand that his "anxiety and feelings of inadequacy" were "due to some childish conflict and have very little to do with reality, at the present time."[70]

There was also not a little self-pity mixed with the anger and depression, at least in the early days. Peale thought he was being "mistreated and persecuted" and classified as a bigot simply because he did not "vote with the Catholics."[71] The election of Kennedy, although he predicted it, exacerbated all his anxieties. He exploded to friends: "I think Kennedy is a jerk!"[72] In his opinion, "Protestant America got its death blow on November 8th." As far as he was concerned, "it makes no difference to me who runs for what in the future—I am fed up with it and through with it."[73]

A Lapse in Memory

Eventually, of course, Peale did pull back from the edge. In the process of focusing his energies on a new book, he achieved an emotional breakthrough of sorts, accompanied by helpful insights into his inner self. The book became known as *The Tough-Minded Optimist*, and it revealed a new dimension to his usual upbeat formula. It also represented the fullest published expression of Peale's populist critique of the mainstream church. The final section was his judgment on the Protestant liberal establishment, which he held responsible for the decline in the national influence of religion. In the process of writing the book, he controlled his anger, came to terms with his recent mistakes, and more important, accepted the fact that there was a price to pay for taking controversial actions.

To his great credit, Peale did not seek to shift blame to the other conference leaders, many of whom were more involved with arranging for the session than he had been. The most notable figure to remain in the background through the heat of the controversy was Billy Graham, though it was, in fact, he who had hosted the planning session in Montreux. Peale never attempted to shift or share culpability with Graham, and Graham never publicly came forward to explain his role.

Dr. J. Elwin Wright, the prime mover behind the creation of the Citizens for Religious Freedom Committee and a man who had Graham's ear, presumably advised Graham to keep his distance from the Washington meeting and the ensuing controversy because of their potential for adversely affecting future crusades in Catholic countries. Wright, seventy years old in 1960 and a Congregational minister who had been in executive church work for all his career, was a committed evangelical who worshiped at Ockenga's Park Street Church in Boston. His alleged "concern over the possibility of a Roman Catholic in the White House" had prompted him to contact the Reverend Donald Gill of the National Association of Evangelicals to learn what actions could be taken.[74]

When a disgruntled evangelical complained to Wright about Peale's withdrawal from the group in the wake of the public uproar, Wright supported Peale's decision. Pleased with Peale's seemingly recent embrace of the evangelical movement, Wright worried that Peale's further identification with the conference could limit his effectiveness.[75] It was precisely the fear of this eventuality, Wright noted, that led him to urge Graham "*not* to become identified with this controversy in any way" because it could affect his access to the media and, more important, affect his ability to "get visas to any Catholic dominated country (in which his ministry could be so helpful)."[76]

Peale's good friend Dan Poling was not quite so willing to relieve Graham of all responsibility. Poling told Graham that his own involvement in the Washington meeting was prompted by his belief in "your own leadership in the arrangements."[77] It had been his intention, explained Poling, to limit his role in the campaign to writing editorials until he learned that "such men as Doctors Peale and Ockenga and you" were to be in Washington.[78]

When Graham responded, explaining his understanding of the arrangements for the meeting, he shifted the burden to Peale, indicating that it was Peale who first contacted him. As he explained to Poling:

> He [Peale] sent a telegram stating that he was concerned about the election and would like to talk to me. I responded by saying that we were having an evangelical conference in Montreux and that we were going to have a meeting of the American delegates on a particular afternoon. I invited him to attend.[79]

At the meeting, Graham pointed out, Peale acknowledged that his forthright position could be costly, that he could lose friends, his newspaper column, and the influence of *Guideposts*, but he felt he had to stand his ground. As for himself, Graham said, he told the group privately his views on the issues, but that his preference was to work "behind the scenes" rather than in the public domain. He knew nothing of the Washington meeting, Graham explained, until the publicity broke, and only much later did he learn of Dr. Wright's plan to keep him in the dark so he would not become involved. According to Graham, Nixon had told him he preferred having clergymen work on getting out the vote rather than becoming engaged in the campaign. Had he come out "wholeheartedly" for Nixon, said Graham, the result for the Vice President would have been "untold harm to his chances."[80] He explained to Poling that he had tried unsuccessfully to reach Peale to talk about such matters, for he believed Peale had "made a mountain of a molehill" in estimating the damage done to his reputation.[81]

Poling found that Graham's explanation, which seemed to place all the blame for the conference planning on Peale, did not square with his own sense of the events, with what he remembered Graham himself saying the previous spring—that he would "stump for Nixon" and cover the country in his support—and with what other participants at the Montreux meeting told him.[82] Admitting to a Quaker "concern" about the seeming disparity, and emboldened, he confessed, by the realization that he was in "the sunset" of his life, Poling confronted Graham with the conflicting evidence. He told Graham that he learned in conversations with those who attended the conference in Switzerland that they had gone to the Washington meeting specifically at Graham's urging and that they recalled Graham himself offering details about the "moral character of one of the Presidential candidates."[83] Graham and Poling continued to talk and correspond about the matter until, finally, nearly seven months later, Graham admitted to a lapse in memory, to being "red-faced, humble and apologetic," in a letter to "My beloved Dan": "Please believe me [he wrote], I did not deliberately misrepresent the situation. It was completely a slip of memory. I cannot possibly understand why it happened."[84]

As he now recalled, he had actually encouraged Peale to attend the Washington meeting. It was only because he had learned after the meeting was over from Elwin Wright and L. Nelson Bell that they had purposely kept information about the conference from him that he concluded himself that he "had not even heard of the meeting." Sounding contrite, he offered his apologies.[85] Graham's letter thus made it appear that two leaders of the evangelical movement, himself and Peale, were adversely affected by this last-ditch effort to prevent the unthinkable, the election of a Roman Catholic President. Peale's suffering was public and sensational, Graham's private and personal. In the last days of the campaign, Graham declared publicly for Nixon, informing Kennedy he did so out of personal friendship for the Vice President and not for religious reasons.[86]

Nixon Redux

Peale's recovery, psychologically and professionally, from the effects of the controversy was slow. At least two developments facilitated his rehabilitation; one was producing the new book, *The Tough-Minded Optimist*, and the other was the acceptance of the fact that he had not contributed to the defeat of Nixon. Two weeks after Kennedy's election, when Ralph McGill of the *Atlanta Journal and Constitution* suggested in a column that Peale's performance may actually have made such an outcome inevitable, Peale responded as if wounded all over again: "If it were true [he wrote to McGill] that I had ever destroyed any man, especially so fine a gentleman as Mr. Nixon, I would feel my life a failure and would regard it as my duty to consider retiring from the ministry and from writing and speaking."[87] If guilt is what he was feeling, much of it was assuaged when Nixon, as one of his last acts in office as Vice President, wrote him, for the first time a "Dear Norman" letter, saying "The responsibility for losing it [the election] is one that I and I alone must bear."[88] Nixon indicated that mutual friends had told him that Peale was troubled by the published claims (perhaps such as McGill's) that the Washington statement of the previous September had affected the election results. If that was the case, said the Vice President, "you should banish such thoughts from your head once and for all."[89] He suggested that the two families, the Nixons and the Peales, renew their friendship when he returned to New York.

Peale seemed almost redeemed by Nixon's words, even allowing Nixon's reference to Peale's church as Brick Church rather than Marble Church to pass unacknowledged. He confessed to Nixon that he had been "literally broken-hearted by the turn of events," by the possibility that "I, who love you and believed in you so deeply, might have contributed in any way to the defeat."[90] He had a hard time, he said, forgiving himself. But he continued to believe in the nobility of the man, which in defeat had put "Richard Nixon in the galaxy of the truly great men America has produced," and would ensure that "your greatest days are ahead," he said prophetically.[91] The two men remained in contact and on such friendly terms that when Nixon ran for governor of California a mutual friend urged Peale to come West to promote his election. In refusing, he expressed the fear that his presence probably would not help and, in fact, "might hurt."[92]

Nixon's reassurance gave Peale the comfort he needed and that no one else could have provided. Peale had identified with Nixon on a personal basis, as earlier he had related to Dewey, and he grieved this defeat more bitterly still, the more so because he felt that he had contributed to the final outcome. He experienced it as a moral indictment. It was as if in an earlier time, a warrior for the cause of temperance, he had been discovered selling bootleg gin. Nixon's concern was restorative.

The effort it required to produce the new book had the same effect. It appeared in bookstores a year after the election, in November 1961, just as Peale's career was beginning to rebound from the pounding it had taken. The text was essentially vintage Peale, with the toughness in *The Tough-Minded Optimist* contained in an attack on the liberal Protestant clergy. He indicted the mainstream church for many of the ills in the culture, from the decline in morals to the absence of "religious freedom" to the erosion of national strength.[93]

John Bennett and the Union Mystique

For reasons not clear, though perhaps subconsciously hostile, Peale sent a copy of the book to John Bennett, a person who in his mind was one of the major players in Protestantism's decline. After writing Peale a rather generous note, thanking him for the book and going on to explain that he had never applied the "Protestant underworld" designation specifically to him (although he admitted to connecting him to the term "bigotry" in an "impersonal" way), Bennett felt compelled to follow up with a second, quite lengthy letter to his New York neighbor. He had had a chance to read the book, he said, and he wanted to respond to some of the allegations Peale made about the church and its leadership. It would have been easy for Bennett to assume that the charges were intended for him and his close associates, people like Reinhold Niebuhr and Paul Tillich.

Bennett's prose was artful. After suggesting that their differences were so historic that perhaps he and Peale should simply agree to disagree, he then went on to examine some of these differences. His theology, he explained, was not tied exclusively to "social action" but was premised on the belief that "on the deepest level the individual is the unit of decision and commitment."[94] When Bennett turned his attention to Peale's own message, he observed "that you have been attacked by theologians and ministers who believe that you give over-simplified answers which . . . obscure the Gospel more than they reveal its meaning. But this does not cancel the healing power of what you do for many whom you do reach."[95] He defended himself against the charge of "leftist," explaining that he and his colleagues were "moderate, pragmatic rather than doctrinaire believers" in any specific ideology, who nevertheless regarded Peale's version of "economic individualism" as "one-sided" and inhibitory when it came to dealing with the nation's collective problems.

He offered a concrete example of an issue he thought focused their differences—nuclear war. Suppose, he asked, all the "changed individuals" Peale helped produce were suddenly faced with the decision of destroying the entire population of another country. The dilemma would then be how to prevent the extension of communism while also preventing a nuclear war, a major issue, said Bennett, "worthy of your attention." If Peale, for whatever reason, chose not to deal with the issue, Bennett argued, then "I do not see why you should discourage others who feel that this problem cannot be dodged."[96]

Peale responded immediately to Bennett's allegations in an almost equally well-worded and lengthy letter. He claimed that he was not sure the two of them did disagree, even in substance, for he read many of the same theologians Bennett endorsed. As for the charge of his own "oversimplifying," which he didn't think was the case, he did it because "I feel that the spiritual illiteracy of the general public is so great" that it requires simple terms and concepts.[97] He denied that he called people "leftist" or "rightist"; when these terms appeared in his book they were quotes taken from others, and he personally disliked their use.[98] Peale admitted that he opposed Bennett's "Red China obsession," but he agreed with his point about nuclear war. Then revealing that even a positive thinker could have a dark and shadowed side, he confessed to believing "there will be a nuclear war." He was also convinced that "some type of communism is bound to dominate mankind increasingly. . . . [I]t is my considered judgment that a modified form of dialectical materialism is probably destined to have its day in human affairs."[99]

Peale was more defensive when he turned to allegations about his own ministry. Preachers like himself, he reminded Bennett, had been "sneered at and treated with contempt because we try to do a simple job helping simple, everyday people live in this unhappy world." The critics were mistaken to assume that "I am simply a fortuitous and Pollyanna character," he said. In fact, he argued, the Protestant faith in America was in "rapid deterioration" and "has very little potency now." He thought it wrong of Bennett to assume preachers should "cover all questions of society," but rather that they should do what they have the talents for.

Further, he said he disliked Bennett's references to Peale "attacks," for until the present book he had never been an attacker: It was true, however, that after being "kicked around by a certain type of religious leader," he was no longer so "nicey-nice" and was willing to put up a fight if it seemed called for. In addition, Peale denied being "the apologist of the business man."[100]

In answering this extraordinary letter, which was by turns dark, cynical, elliptical, and reassuring, Bennett confronted what was for him a revelation about Peale that he found surprising. The liberal professor of social ethics noted, "it is I who am the more optimistic about some of the things concerning which you write." Bennett also observed that it could only be beneficial to Peale's son John for the two older men to air their differences rather than to keep them covered up.

Overall, the letters were near-classic statements about the issues separating the two positions—establishment and dissenting—and the distance between them. Bennett soon after became the president of Union, leaving the seminary administrator and the populist preacher to pursue their separate missions just a few miles apart in Manhattan.

The Aftermath

The Kennedy controversy sharply focused the issues that had characterized Peale's ministry. His religious populism, conservative politics, and social sensitivities were all joined in a confrontation with what he believed were the morally and culturally relativistic values of an establishment contemptuous of people like himself. His challenge was simultaneously theological, political, sociological, psychological, and cultural. And unfortunately for a populist leader, it was no longer the dominant view, even within popular culture. Both the evangelical challenge and the Kennedy election were benchmarks in the acceptance of American cultural pluralism.

Peale's worst fear, that the crisis had done him "irreparable harm," did not materialize, although for months media reporters continued to follow him with questions. Frustrated, he asked the editor of *The Washington Post* to call off his journalists, since having been "effectively silenced" by the persistence of the reporters, he never intended to speak of the event again.[101] Yet latent hostilities could still bubble to the surface, as they did when he preached in Marble Church in April 1961 on the occasion of the fortieth anniversary of his first sermon, again criticizing the "contemporary organized church" for its role in the "depressing decline of Protestantism" and its lack of interest in "saving souls."[102] Significantly, Peale detected a bright spot emerging to fill the loss, and it was to be seen in the "development of creative spiritual cells, including mostly business men." By 1962 *The Philadelphia Inquirer* had restored Peale's column, and although the stipend he received for it was substantially reduced, its reappearance was an important symbolic victory.

Public interest had moved on to other, more critical matters, notably civil rights and the developing war in Vietnam. The tone of Peale's message, with its evangelical-New Thought blend, was preserved in the burgeoning ministry of Peale-protegé Robert Schuller, whose variation on Peale's theme was "possibility thinking." Like Peale, Schuller was also a minister within the Reformed Church in America, and he was as concerned as Peale about our "national spiritual heritage."[103] The election of Nixon in 1968 went a long way toward absorbing Peale's trauma of the past: however, the tumult in the culture made it clear that this was not a restorationist marker for the old Protestant order.

The Nixon-Peale friendship could now safely be a matter of public record. Nixon's younger daughter, Julie, was married by Peale to David Eisenhower in Marble Church shortly after her father's election. Nixon attended worship services in Marble Church on occasion, and Peale was invited periodically to lead services in the White House. In 1969 Peale accepted the President's invitation to visit the soldiers in Vietnam, a trip widely reported in the press. That same year, in a *Wall Street Journal* assessment of Peale's ministry (including a literal attempt to assess his financial worth), the paper observed that although Peale would not take credit for Nixon's comeback, "Mr. Nixon could be responsible for Mr. Peale's."[104] There were some interesting parallels in their similar journeys through crises and vales of tears, although the newspaper agreed that Peale's career had never been eclipsed as Nixon's had been. Peale's most effusive tribute to the President was that he was "the greatest positive thinker of our times.[105]

Slowly, inevitably, Peale returned to the world of political commentary. Older, more subdued, sobered by some trying experiences, he learned to preface all comments to reporters with the demurral about its not being appropriate for members of the clergy to get involved in politics. But it usually proved too difficult a temptation for him to resist. Asked about his political preferences on a 1970 visit to California, he answered that he was a "conservative, but not a hawk." Then, after putting in a good word for friends Nixon and Spiro Agnew, he offered a plug for local candidates Governor Ronald Reagan and Senator George Murphy.[106] His highly publicized friendship with Nixon put him back in the political limelight. Reinhold Niebuhr remarked that he thought it inappropriate for Nixon to invite members of the clergy to the White House to conduct religious services, pointedly observing that it encroached on issues of church-state separation.

Peale stuck by Nixon through his darkest hour, the national trauma leading up to his resignation of the presidency in 1974. In the aftermath of Nixon's re-election, Peale had hailed him as a "peacemaker," citing him for the "leadership and direction [that] so substantially fortified the moral fiber of America and strengthened the future of our society."[107] As the Watergate crisis deepened, Peale joined the committee called "Americans for the Presidency," which asked people to write to their members of Congress, as well as members of the House Judiciary Committee, to let the President "get on with the vital business of insuring our strong leadership in the world." In becoming a public sponsor, he obviously compromised his own pledge to stay out of all forms of political participation.

Several years after Nixon's forced resignation, when a reporter asked Peale about his relationship to the President during that trying period, he answered that Nixon was a private person who had not sought his advice.[108] In responding to an inquiry about how Watergate could have happened, Peale said: "I don't believe it would be fitting for me to comment on what any member of my church does. . . . I feel a sense of compassion, or regret . . . it's too bad. But I'm not one to throw stones and I never criticize from the pulpit."[109]

Peale's political presence was understandably less visible during the Carter years, but with the election of Ronald Reagan it appeared as if the populist political wheel had turned full circle, and he was again drawn into affairs in Washington. Reagan was in so many symbolic ways a throwback to the old order that Peale delighted in having a kindred spirit occupying the White House again. Reagan returned the confidence. In 1984 Reagan awarded him the Presidential Medal of Freedom.

Notes

1. In the files on the Kennedy campaign in the NVP Manuscript Collection, there is a collection of pro and con letters regarding Peale's participation in the event. The majority of letters favor Peale's role, but not by much—by about a 5 to 4 margin.
2. NVP to the Rev. Phillip Ellman, Portland, Oregon, May 7, 1948; George H. Sibley to NVP, July 7, 1948; NVP Ms. Coll. In Sibley's letter he told Peale that he had mentioned to Dewey the help Peale was giving him in the Oregon campaign.
3. NVP to Dr. Elmer Ellsworth Helms, Dec. 31, 1948; NVP Ms. Coll.
4. *New York Herald Tribune*, November 1, 1948.
5. NVP to Dear Fellow-Minister, Oct. 20, 1950; NVP Ms. Coll.
6. NVP to Mayor Vincent R. Impellitteri, Dec. 17, 1951; NVP Ms. Coll.
7. Statement of NVP at Marble Collegiate Church, Nov. 29, 1950; NVP Ms. Coll.
8. *The New York Post*, October 7, 1955.
9. Ibid.
10. NVP to Dr. Emory S. Bucke, *Zion's Herald*, Oct. 21, 1947; NVP Ms. Coll.

11. *Charleston Daily Mail*, April 14, 1960.

12. Ibid.

13. NVP to Mrs. A. B. Thomas, Jr., Charleston, W. Va., Apr. 14, 1960; NVP Ms. Coll.

14. Circular letter from Frank Sayre to "Fellow Pastors in Christ," Apr. 1960; NVP Ms. Coll.

15. NVP to Frank Sayre, Apr. 14, 1960; NVP Ms. Coll.

16. *The World Telegram and Sun*, March 26, 1960.

17. NVP to Robert Kennedy et al., May 17, 1960; NVP Ms. Coll.

18. Typescript, May 10, 1960; NVP Ms. Coll. Excerpts also appeared on May 9 in *The New York Times*, which attracted considerable attention.

19. Harold John Ockenga, "Religion, Politics and the Presidency," June 5, 1960; L. Nelson Bell, "Protestant Distinctives and the American Crisis," Aug. 21, 1960; NVP Ms. Coll.

20. Daniel Poling to NVP, Aug. 2, 1960; NVP Ms. Coll., Box 327.

21. RSP to Kurt Volk, Bridgeport, CT, Aug. 19, 1960; NVP Ms. Coll.

22. Ibid.

23. Ibid.

24. NVP, "What a Protestant Should Do Today," typescript, NVP Ms. Coll. This is the only copy of this statement I have seen. It is possible that it was never printed and thus never circulated. It is significant for revealing Peale's thinking about the issues in August 1960.

25. Citizens for Eisenhower-Nixon to "Dear Fellow Clergyman," n.d., 1952, typescript copy; NVP Ms. Coll. Peale was on the board of the Protestant Clergyman's Committee for Citizens for Eisenhower-Nixon, and he was responsible for sending out the "Dear Fellow Clergyman" letter.

26. Richard Nixon to NVP, May 13, 1954; NVP Ms. Coll.

27. Hannah Nixon to NVP, Mar. 28, 1960; NVP Ms. Coll.

28. NVP to Richard M. Nixon, Apr. 5, 1960; NVP Ms. Coll.

29. Richard Nixon to NVP, Apr. 15, 1960; NVP Ms. Coll.

30. NVP to The Honorable Richard M. Nixon, June 23, 1960; Richard Nixon to NVP, July 15, 1960; NVP Ms. Coll.

31. NVP to "Dear ______:", n.d., typescript copy; NVP Ms. Coll.

32. Richard Nixon to NVP, July 11, 1960; NVP Ms. Coll., Box 167.

33. NVP to The Honorable Richard M. Nixon, Aug. 1, 1960; NVP Ms. Coll., Box 167. This is an unusually long letter, three pages, to have been sent from overseas. Peale usually sent his letters when he was away to his church secretary, who typed them and sent them out.

34. Mary Creighton (secretary to Dr. Peale) to Dr. Daniel Poling, Aug. 24, 1960; Daniel Poling to Mary Creighton, Aug. 25, 1960; NVP Ms. Coll. Poling's letter said, in part, "Sept 7th is already fixed as a date for a larger conference in Washington, D.C. which Norman is to attend."

35. Office memo of Mary Creighton, Aug. 29, 1960, typescript; NVP Ms. Coll. The memo comments: "Dr. Peale knew this meeting was in prospect, as he attended a conference in Switzerland of clergymen and other leaders brought together by Billy Graham." In the cable that followed the memo, Peale was asked if he could preside at the Washington meeting on September 7 at 9:30. Peale's cable reply, also dated the 29th, says, "Yes September 7th Peale."

36. See note 34 above, Poling's letter of August 25, which notes the September 7 date.

37. Donald H. Gill to NVP, Aug. 29, 1960; NVP Ms. Coll.

38. An invitation and a copy of the program, with Peale's instructions to himself about conducting the meeting, are in the NVP Manuscript Collection.

39. According to *Time* magazine (Sept. 19, 1960), the reporters were John J. Lindsay of *The Washington Post and Times-Herald* and Bonnie Angelo of Long Island's *Newsday.* They reported Peale as saying, "Our American culture is at stake. I don't say it won't survive, but it won't be what it was."

40. Typescript copy, NVP Ms. Coll. Also *The New York Times*, September 8, 1960. The meeting was front page news in all the papers.

41. *The New York Times*, ibid.

42. *The New York Times*, September 11, 1960.

43. *The New York Times*, September 16, 1960. Peale prepared a formal statement explaining his role and his departure. The statement appeared in many places.

44. Typescript, to "Dear Friends," Sept. 15, 1960; NVP Ms. Coll.; *The New York Times*, September 16, 1960. The statement said in part:

> While in Europe I received an invitation to attend a Study Conference on Religion and Freedom scheduled for Washington on September 7. Beyond the names of a few people who were to be present, I had no information on the meeting itself, but I had respect for those whose names were given to me.
>
> I arrived in New York from Europe on the afternoon of September 6, and therefore had no opportunity to secure further information on the meeting. Had I not agreed to attend a Spiritual Retreat of another group, which included some of our own church members, on September 8, in Washington, I probably would not have attended the September 7 Religion and Freedom meeting.

45. Ibid.

46. *The New York Times*, September 19, 1960.

47. *Christianity and Crisis*, September 19, 1960.

48. Ibid.

49. Ibid.

50. Daniel Poling, "Statement of Sunday Night," September 18, 1960; NVP Ms. Coll.

51. John C. Bennett to NVP, Jan. 3, 1962; letter from personal file of Dr. Bennett. In an interview with Bennett in Claremont, California, in October 1987, he said that the president of Union at the time, Henry Pitney Van Dusen, who had been converted during a Billy Sunday campaign, thought that the statement that Bennett and Niebuhr issued at their press conference about "blind prejudice" was one of the worst things to come out of the Seminary.

52. *The New York Times*, September 19, 1960; *The New York Herald Tribune*, September 16, 1960.

53. *New York Herald Tribune*, September 16, 1960.

54. *Time*, September 26, 1960.

55. Ben Hibbs, editor, *The Saturday Evening Post*, to NVP, Oct. 26, 1960; NVP Ms. Coll.

56. W. W. Forster, editor, *The Pittsburgh Press*, to NVP, Sept. 28, 1960; NVP Ms. Coll.

57. John S. Knight, publisher, *The Detroit Free Press*, to NVP, Nov. 28, 1960; NVP Ms. Coll.

58. *The New Republic*, September 19, 1960.

59. NVP to James Fifield, Sept. 28, 1960; NVP Ms. Coll.

60. NVP to Len LeSourd, *Guideposts*, Sept. 28, 1960; NVP Ms. Coll.

61. RSP to Robert Hall, The Hall Syndicate, Oct. 14, 1960, NVP Ms. Coll.

62. Charles B. Mills to NVP, Sept. 17, 1960; NVP to Charles B. Mills, Sept. 21, 1960; NVP Ms. Coll. An old college friend, Mills thought that the conference had hurt Republican interests.

63. NVP to Phyllis Alexander, The World's Work, Kingswood, England, Oct. 25, 1960; NVP Ms. Coll. He observed, "I think that Kennedy has been taking the attitude that anybody who does not vote for him is a bigot, which of course, is quite a distorted concept."

64. NVP to John Sherrill, *Guideposts*, Sept. 28, 1960; NVP Ms. Coll.

65. NVP to Gerald Dickler, Oct. 25, 1960; NVP Ms. Coll.

66. NVP to Dr. J. Richard Sneed, First Methodist Church, Los Angeles, Jan. 10, 1961; NVP Ms. Coll.

67. NVP to Dr. Smiley Blanton, Jan. 2, 1962; NVP Ms. Coll.

68. Ibid.

69. Smiley Blanton to NVP, Jan. 9, 1962; NVP Ms. Coll.

70. Ibid.

71. NVP to Senator Wilton E. Hall, Anderson, SC, Dec. 7, 1960; NVP Ms. Coll.

72. NVP to Helen and Horace Johnson, Nov. 15, 1960; NVP Ms. Coll.

73. Ibid.

74. *The New York Times*, September 16, 1960.

75. J. Elwin Wright to Luke H. Proud, Los Angeles, Sept. 21, 1960; NVP Ms. Coll.

76. Ibid.

77. Daniel A. Poling to Dr. Billy Graham, Oct. 24, 1960; NVP Ms. Coll. The letter has a handwritten "Here it is" comment from Poling to Peale.

78. Ibid.

79. Billy Graham to Dr. Daniel A. Poling, Oct. 26, 1960; NVP Ms. Coll. Poling obviously shared a copy of the letter with Peale.

80. Ibid.

81. Ibid.

82. Daniel A. Poling to Billy Graham, Nov. 1, 1960; Daniel A. Poling to NVP, Feb. 16, 1961; NVP Ms. Coll. Poling wrote in his letter to Peale that he thought it was wrong of Graham to place the "entire onus of what happened in Switzerland on your shoulders." He had spoken with Bob Pearce of World Vision and based on that conversation had concluded that "what Billy writes is just about unthinkable as to the facts of the case."

83. Daniel A. Poling to Billy Graham, Feb. 16, 1961; NVP Ms. Coll.

84. Billy Graham to Daniel Poling, June 8, 1961; NVP Ms. Coll., Box 327.

85. Ibid.

86. "The Capitol Merry-Go-Round" column in *The White Plains Reporter Dispatch*, November 4, 1960. The article pointed out that in Graham's Washington crusade the previous June, he and his wife had been guests at the Nixon home several

times. They had also occupied seats reserved for the Vice President in the Senate visitor's gallery.

87. NVP to Ralph McGill, Nov. 23, 1960; NVP Ms. Coll., Box 165.

88. Richard M. Nixon to NVP, Jan. 18, 1961; NVP Ms. Coll. After addressing the letter "Dear Norman," Nixon said, "As I dictate this letter I am somewhat hesitant to address you by your first name but I feel that I know you so well that you will pardon me for using this personal salutation."

89. Ibid.

90. NVP to the Honorable Richard M. Nixon, Jan. 29, 1961; NVP Ms. Coll.

91. Ibid.

92. NVP to Floyd L. McElroy, Woodside, California, Sept. 18, 1962; NVP Ms. Coll., Box 165:

93. Peale, *The Tough-Minded Optimist* (New York: Prentice-Hall, 1961).

94. John C. Bennett to NVP, Jan. 5, 1962; copy from Bennett personal files.

95. Ibid.

96. Ibid. In an interview twenty-five years after this exchange of letters, Bennett spoke of Peale in a quiet, neutral way. He said he recalled little of those heated events in the sixties. His impressions of Graham were more generous and fulsome than they were of Peale. Graham, he pointed out, had taken an antinuclear position and tried to be ecumenical. Interview, Claremont, California, October 16, 1987.

97. NVP to Dr. John C. Bennett, Jan. 9, 1962; copies in NVP Ms. Coll. and Bennett files. During this exchange of letters, Bennett and Peale lived about two miles apart in New York City, Bennett on West 116 Street and Peale on Fifth Avenue and Central Park. Theologically and politically, of course, much greater distance separated them. Fortunately for the researcher, they did not carry on the conversations in person or on the telephone.

98. Bennett was probably aware, however, of a quote attributed to Peale himself that had appeared two months earlier in *The San Francisco Chronicle*, which argued that "a left-winger like John Bennett of the Union Theological Seminary" had gone too far in emphasizing the social doctrines of the church. *San Francisco Chronicle*, November 19, 1961.

99. NVP to Bennett, Jan. 9, 1962. See note 97.

100. Ibid.

101. NVP to editor, *The Washington Post*, Mar. 1, 1961; NVP Ms. Coll. The managing editor, Alfred Friendly, responded that he was mystified by the charge of silencing, for that was not their intention. "We abhor silence," he noted, "as nature abhors a vacuum." Alfred Friendly to NVP, Mar. 1, 1961; NVP Ms. Coll.

102. Press release of summary of sermon, typescript, Apr. 4, 1961; NVP Ms. Coll.

103. Robert Schuller to Daniel Poling, Apr. 10, 1961; NVP Ms. Coll. In this long and lively letter, Schuller wrote: "[S]omething big is about to happen!" He envisioned a sort of national council of evangelicals—to include Catholics like Fulton Sheen, and a rabbi, he hoped—to preserve the faith and the nation. Peale and Schuller were good friends, speaking frequently from each other's pulpit.

104. *The Wall Street Journal*, May 7, 1969. The lengthy article attempted to show both sides of the controversial Peale, although it was more favorable than not.

About Peale's wealth it said: "Mr. Peale won't say how much he earns during a year or how much he is worth. But he surely is as wealthy as some of his executive friends. He is said to get $1,500 for each of his 100 or so speeches a year. He probably gets a minimum of $3,500 for each *Reader's Digest* article. His newspaper column probably brings in at least $50,000. And he probably gets a substantial salary from Marble Collegiate, which has an annual budget of $565,000 and a payroll of 35. People in the area say his Pawling estate could be sold for as much as $500,000.

"Mr. Peale won't say how he spends or gives away his money, except to say that he frequently turns his speaker's fees over to charity, especially if he is speaking at another Dutch Reformed Church."

These estimates by the reporter are probably quite wide of the mark. While Peale's salary from the church was comparatively high as ministerial salaries go, it was probably not much out of line with those of other highly visible New York ministers. His attractive Fifth Avenue apartment was a bonus, surely. But the real source of his wealth was his book royalties. *The Power of Positive Thinking* ensured his status as a millionaire within two years of its appearance in 1952. Ruth wisely invested their money from the beginning, which, combined with their unostentatious life style, left considerable surplus to be reinvested.

105. *San Francisco Chronicle*, November 3, 1970.
106. *San Francisco Examiner*, November 2, 1970.
107. *The New York Times*, November 27, 1972.
108. *Sarasota Herald Tribune*, October 22, 1977.
109. Ibid.

CHAPTER 8

The Third Disestablishment

Standing on the bare ground,—my head bathed by the blithe air, and uplifted into infinite space,—all mean egotism vanishes. I become a transparent eyeball. I am nothing. I see all. The currents of the Universal Being circulate through me; I am part or particle of God.

Emerson
Essay on Nature
1836

Frustrated and embittered by a second major confrontation with the media, most recently over politics as earlier it had been over theology, Peale finally resolved to withdraw from all activities that might involve him in larger communities of the church and civic life. While hardly ignoring the public contemporary scene, he committed his energies to developing his personal ministry. From 1960 until his 1984 retirement from Marble Church, his major concern was enlarging the ministry of the Foundation for Christian Living, reserving his time largely for audiences he knew to be congenial to his work. Essentially what that meant was expanding his communications network of books, radio programs, articles, and various other publications and cultivating his contacts in the business community. He estimated that he spent more than a third of his time with this latter group.[1]

Peale was always a person of great paradox, and never more so than during these years. At a time in life when many Americans accepted the ease associated with senior citizen's status, Peale continued to fly around the country, racing from one speaking commitment to another. And although he had become a fixture in American cultural lore, his trademark if not his name a household term, he persisted in trying new measures for reaching larger audiences. He had become quite fully alienated from the institutional church, yet, paradoxically, he responded positively when offered two prestigious and essentially honorific positions in the church hierarchy. In 1965 he survived a contested election to serve the first of four annual terms as president of the Protestant Council of the City of New York.[2] And in the

1969–1970 term he was elected president of the General Synod of the Reformed Church in America, achieving a belated recognition from the denomination to which he had directed large contributions of members and money. Despite holding these high offices, however, he continued his bold criticisms of church leadership in sermons and publications while describing ecumenical programs as failed efforts to "merge weakness." But to zealots who perceived encouraging evidence from these indictments of a fellow enemy of the National Council of Churches, Peale responded with little support, repeatedly endorsing both the loyalty and undoubted good will of the Council's officials.[3]

Peale's paradoxical behavior in his public church life had its counterpart in his theological outlook. He admitted that he was simultaneously both a modernist and a fundamentalist. Claiming to have preached basically "the same sermon in this pulpit every Sunday for thirty years," he told a Marble congregation that allegations that he was both a modernist and a fundamentalist were true.[4] The admission seemingly created no inner conflict or cognitive dissonance for him, such as might result from a sense of a clash in values. And yet in a curious and timely way this dual identity served him well in his ministry. On one level, it enabled him to appeal to growing constituencies on both ends of the religious continuum—the secularly inclined New Age believers and the born-again evangelicals. These were the new recruits to personal religion helping to "restructure" American religion in the postwar decades. This double identity also allowed Peale to offer himself as a model for the culture, a kind of work in process, whose self-confidence and personal esteem emerged from dialectical tensions operating in his own life. It might have been a relevant example to some of those engaged in the much-analyzed quest for meaning during the troubled sixties.

A popular description of the religious scene in the two post-Kennedy decades is of a time when "the middle dropped out," or in a more historical sense to name it a "third disestablishment" of Protestantism.[5] It was third in a sequence that began with the disestablishment of religion (read Protestantism) with the American Revolution and entered another stage with the disruption of Protestant cultural hegemony in the 1920s and thirties. Then during the decade of the sixties the witness of the moderate middle in Protestantism was almost overwhelmed by a chorus of new voices from the religious right and left. The center seemed unable to hold.

Scores of new religious bodies holding unconventional views appeared on one end of the religious spectrum, while on the other there were conservative, evangelical groups enjoying fresh popularity among many middle-class Americans.[6] What held these two poles together in a kind of holy tension, apart from a shared opposition to the priorities of the mainline liberal churches, was a culturally attuned regard for experiential religion and a renewed emphasis on the needs of individuals. A need Peale had long understood, it had become the defining quality of contemporary religion. As Robert Wuthnow examined the altered religious landscape in these years, he discovered that a vital desire for personal religion was pervasive, as people searched for a religion that would "champion the needs of the individual as opposed to the rising specter of vast impersonal institutions."[7]

Until recently, those who studied this phenomenon of the cultural re-emergence of personal religion and privatized worship located its roots only in the countercultural developments of the sixties, in particular the student movement.[8] Peale had pioneered in the territory at least a decade earlier and had already marked out new directions in the cultivation of individual consciousness and the qualities of the "more" life. His hybrid message of conservative politics and harmonial-New Thought theology was ideally positioned to win supporters on the New Age left and the evangelical right. Because of his politics and views on social morality he was attractive to the evangelicals, and because of his positive thinking mysticism he was appealing to countercultural seekers after forms of personal fulfillment. It is not surprising that Peale's personal ministry grew at a time when mainline Protestant churches declined.

The student movement, for example, argued that bureaucracies and mainline institutions like the church had for too long been perpetuating wrong values while ignoring the disenfranchised. In fact, Peale's whole ministry in the postwar years had been predicated on a version of popular religion inherently anti-elitist and anti-institutional. Indeed, as Nathan Hatch has shown in his study of religion in the New Republic, populist religion in America has historically been characterized by expressions of anticlericalism, antiprivilege, and anti-theological orthodoxy, in the name of celebrating the ordinary and the vernacular; Peale's popular religion was no exception to this model.[9] Religious populism is by definition a movement at odds with the genteel or high-culture tradition, hospitable instead to the beliefs and pragmatic desires of "ordinary" people, the audience Peale so assiduously cultivated. As both a status revolution and an alternate belief structure, Peale's populist crusade, shaped around a kind of evangelical mysticism, had already portended by 1960 a countercultural model.

His syncretic theology was sufficiently plastic to find appeal with a diverse, even polarized, audience. It utilized traditional biblical language to explain both the power of individual mind—in a way mystics could approve—and the need for a personal, experiential encounter with the Divine Presence—as evangelicals might desire. Since Peale had superimposed harmonial-New Thought beliefs on his evangelical base, as had most of his followers, he himself was more thoroughly conservative evangelical than he was mystically metaphysical. But it was the subjective quality of his message, with its continuing reference to the formative impact of the power of mind, that was the defining element in his ministry and his link to that re-emergent tendency in the culture. As Allan Broadhurst noted in his study of Peale sermons, it was this emphasis on personally guided positive thinking that distinguished a Peale sermon from those of his traditional Protestant contemporaries.[10]

After 1960, and probably related to his feeling that his frequent guest appearances before business groups required a professional delivery, Peale's public speaking—his own term for both his preaching and lecturing—developed a more polished, finished style. A Peale performance was memorable because of his engaging presentation and the simple content of the message, with advancing age—even into his eighties and nineties—enhancing rather than detracting from his reception. By 1960 the format had become more self-consciously formulaic, marked consistently by reliance on three anecdotes to develop the three points that comprised the talk. With a reputation as a good preacher, he became a superb storyteller, captivating his audiences by the force and imaginative power of the parables he told. These parables were most effective when they depicted the heroic efforts of individuals overcoming crises in their lives. The heroes of these tales were almost always men—at least three quarters of the time—and when he addressed business groups they also frequently assumed the qualities of Horatio Alger types overcoming economic odds. A popular message and a polished presentation became a winning combination for Peale's ministry at a time when his colleagues in mainline churches were beginning to tote up losses.

A Faith for a Time of Crisis

Peale's theology might therefore be called a crisis theology, designed for individuals experiencing personal crises of living. In its unique way, it was a testimony about the chaos and uncertainty of modern life. It was a timely statement about the culture, the sense of homelessness it engendered—psychic, spiritual, and otherwise—and how some individuals sought to cope with it. All of Peale's offerings—sermons, talks, books, *Guideposts*—described people at crossroads' places in their lives. But he advised his audience to welcome situations that threatened as challenges rather than fearing or turning from them. To a Marble Church congregation Peale explained that a crisis was not to be shunned, because "properly used [it] may mean an entrance into a great new world. In any case, the person who expects to live successfully must know how to meet a crisis."[11]

His understanding of crises grew out of a Methodist context of the means to conversion. A sense of crisis, of being brought low, was necessary to the conversion experience that brought salvation. The preacher's obligation was to direct his listeners toward that goal. That was his clear perspective on the duty of the preacher when, in the early 1940s, teaching a homiletics course at the Reformed Church seminary in New Brunswick, New Jersey, he urged his students always to "Preach for a Verdict."[12] Some of the moral urgency in that evangelical sense of crisis softened considerably in his later ministry as he focused less on moral issues and more on matters of daily living: health, family, finance, techniques of personal religion. Yet an evangelical awareness remained for him personally the starting point for understanding the role of crisis: Crises meant opportunity. It was an ironic message for an uncertain time. Acceptance of crisis, as a personal and collective reality, was a necessary first step. The irony in Peale's message was to see modern crises—whether psychic homelessness or physical illness—as challenges.

His theme of crises and how to deal with them was not necessarily fresh evidence that his modern supporters were softer or emotionally less durable than their nineteenth-century forbears, nor was it an indicator of their narcissism. Rather, it seemed that psychic stresses were keener, more enduring, and more ubiquitous, personal futures more tentative and less predictable. These nervous feelings were frequently combined with heightened positive expectations, which made it less likely that people would accept what once was endured, ignored, or salved away. In an expansive postwar economy, seductively suggesting limitless possibility, opportunity co-existed with fearful uncertainty. Nor were the old sources of comfort and support in church and community available to sustain as they once were.

To manage these crises of modern living, Peale believed, required powerful inner resources. The required strength to manage crises came from what he described as the "Jesus Christ within." The only source of "energy, . . . vitality, . . . life," it was unlikely to be found in the dead "formalism" of "nominal Christianity."[13] Increasingly after 1960, Peale's presentations were often word pictures contrasting these two opposing forces, one the vital inner life of personal religion and the other the stultifying intellectual quality of formal religion. Each sermon was a new opportunity to describe the differences between the two experiences: One depended on the internal, energizing power of the Jesus Christ within; and the other relied on the "dead formalism" of a "bureaucratized" church. As theme and counterpoint, it ran through virtually all his sermons in the 1960s and 1970s.

Peale's sermons became occasions for giving imaginative life to this dynamic inner force, and as a gifted storyteller he did this to best effect in compelling anecdotes. In a 1964 sermon at Marble Church, he employed a particularly vivid anecdote to help his listeners get a sense of the awesome inner power available to them, as to the boy in his story.

A small boy, Peale said, once approached an ancient philosopher with a challenging request.

> "Master," said the boy, "I want knowledge."
> The philosopher asked, "How much do you want knowledge?"
> "Oh," the boy said, "I *really* want it."
> "Well," remarked the sage, "to get knowledge you have to want it as you want nothing else." And he continued, "You just come with me, will you." He took the boy to the seashore, where together they waded out until they were in

> water up to the boy's chin. Then the philosopher put his hand on the boy's shoulder and shoved him under the water. For a long moment he held him there kicking and squirming and struggling to be released. Finally, just before the drowning point, he let him up. Back on the beach the philosopher asked, "When you were under the water what did you want more than anything else in the world?"
>
> "Master, the thing I wanted more than anything was air," the boy answered.
>
> "Well," said the older man, "when you want knowledge as you then wanted air you will get knowledge."[14]

But there was something even more important than just the reality and magnitude of this inner power. That was its ability in other contexts, when properly channeled and focused, to be redemptive, to produce life-changing, in some cases, life-saving, results.

In another even more gripping, more dramatic account, Peale told his Marble Church listeners of this other almost unbelievable, transformative power of the Divine Presence within. The illustration, Peale alerted his congregation, came from a story in the *Des Moines Register*. It told of a young, twenty-eight-year-old pilot who had guided his small pontoon plane onto the frozen surface of an isolated lake in northwestern Ontario. Intent on his mission, the pilot left the cockpit, the propeller of the plane still turning, to scout the area. But suddenly he slipped, and hitting his head on one of the pontoons, fell unconscious into the icy water. There was no one for miles around. Apparently revived by the cold water, the young man came to and tried to pull himself from the water by grabbing one of the pontoons. "Then," said Peale, "he discovered to his horror that the propeller had cut his right arm off slightly below the shoulder and he was bleeding profusely."

"So there he is," he continued,

> far from civilization, in a lake, arm cut off, his lifeblood coursing out. What would you do? Worry? That sure would help you a great deal, wouldn't it! What did *he* do? He prayed. And he got an answer. He managed to pull himself up onto the pontoon and into the cockpit. He succeeded in fastening a tourniquet around the bleeding stump. He did what was necessary to lift the plane off the water into a beautiful takeoff. Praying all the while that he would not black out, he flew it 15 miles, landed on another lake, where another pilot took over and flew him to Port Arthur and got him to the hospital. His life was saved, even though his arm was gone.[15]

The cynical and the seeker surely assigned these stories different meanings. Although skeptics might credit the pilot's rescue to extraordinary good luck, a surplus of adrenaline, and the analgesic effect of the icy water and find Peale's theological application irrelevant if not banal, both evangelicals and New Age believers could find in the parables the evidence they sought for a commitment to personal religion.

In these years of his fully developed personal ministry, that is, between

1960 and his 1984 retirement, Peale's sermons developed the theme of the transformative power of personal religion by reiterating three basic themes. Highlighted in his writings as well, these themes were health, personal achievement or success, and the value of experiential rather than formal religion. All three, attuned to the culture, took the individual as their starting place and were potentially appealing to both New Age adherents and New Evangelicals. There is no effective way to measure the results this produced. Numerical data measuring growth, always a problematic way of determining effectiveness, revealed generally steady increases in Peale's audience, while the demographic profile of a "typical" supporter remained the same.

Positive Thinking and Holistic Healing

As Catherine Albanese observed in a discussion of the connection between alternate healing and nineteenth-century spirituality, "Nature's law was a law about how the mind shaped the body, a brief for how metaphysic gave birth to physic." What that meant was that disease was no longer "the result of God's punishment or a test by God to help sanctify the virtuous further still. Rather, it was one's own decision; and saving grace, likewise, would come as the inevitable result of one's initiative."[16] One had the right, indeed the responsibility, to control one's own health. An alternate view of healing, of healing as a process corresponding to the laws of holy nature, formed an integral part of the mental science religious perspective.

Forms of natural, or personal, healing had been wedded to spirituality, of course, since ancient times, in settings that encouraged a view of the priest as a channel for divine healing, or the medical person as the possessor of sacral powers. Healing was natural, and like religion, conformed to nature's laws. Peale was far from being opposed to orthodox medicine, but by emphasizing personal healing through psycho spiritual means, he had aligned himself with this alternate tradition. It had been a consistent theme throughout his entire New York ministry, but it gained fuller attention in his sermons through the fifties and after, just at a time when nontraditional forms of treatment were also finding new favor in the culture.

Personal healing fit naturally with Peale's version of populist religion. Now quite distanced from the institutional church, he grounded his challenge to the mainstream church on an image of a democratic Jesus, one who was a "man-of-the-people" and a grass-roots leader who opposed the bureaucratized religion of the state.[17] Peale's intention was not to create an activist model of social protest, such as might be appealing to sixties' student groups, but rather to establish a populist social context for the dynamic "Jesus Christ within." Acknowledging this inner power, individuals could overcome sin and consequently disease while simultaneously eroding traditional sources of authority. These aspects of his message—personal healing, achievement, autonomy—were not separable parts of his appeal. The ability to transcend disease was undoubtedly the largest power claimed by his version of personal religion, extending as it did to a sense of immortality.

That he was indebted to Christian Science and other branches of the mental science tradition was evident in Peale's willingness to connect sin and sickness. "Sin," he once explained, "lies along the nerve lines of your life, [and] saps its energy," making it impossible for the mind to fulfill its potential if the self is not "wholly rid of sin."[18] Sin was not only synonymous with guilt, hatred, and resentment, it was conducive to physical and mental illness.

Sin in this context was also clearly personal and remediable. If sickness and sin were connected, however, it did not necessarily follow for him that sickness related to the mind exclusively, and therefore outside the jurisdiction of medical science. Important influences in his own life had convinced him that orthodox medicine did indeed have a role to play. In the first place, his own family contained two physicians, his father and his brother Bob. And in the second, he learned much of what he knew about developments in the field from his own Institutes of Religion and Health, which relied heavily on the theories and literature of psychosomatic medicine. From the earliest days of the Marble clinic, Peale had accepted psychosomatic medicine—as hybrid a creation as his own message—as his therapy of choice, for its dependence on the body-mind-spirit concept of holistic healing. His view of personal healing was consequently holistic and not radically singular, dependent on mind alone, as in Christian Science.

He often drew references from the field of psychosomatic medicine for his sermons, finding in them evidence of both the limits of science and the opportunities for individual control. On one occasion he told a Marble Church congregation about a Mayo Clinic report that claimed 25 percent of the cases treated at the clinic were amenable to "the instruments of science," while the other 75 percent were "sending the sickness of their minds and souls into their bodies."[19] He reminded them of the pioneering work in psychosomatic medicine by Dr. Helen Flanders Dunbar. He cited her study, which showed that 50 percent of the people she researched had 85 percent of the accidents, suggesting a certain type person was more prone to accidents than others.[20] Peale told his congregation that there was a practical lesson to be drawn out of the evidence, which was that "If you get mad at your wife, something is sure to happen to you; you will bring it on yourself. She won't have to do a thing but sit and wait for it." At another time Peale found support for holistic healing in the work of Dr. Hans Selye of McGill University—whose research anticipated recent interest in the effects of the brain and mental and emotional functions on the body's immune system—which noted the debilitating consequences for body organs of "stress and emotional disorganization."[21] But even while invoking these new and unconventional concepts about healing, Peale stressed that he was not doctrinaire. The view of holistic healing he supported was quite different from the antimedical stance of mental science advocates; "I am not one of them," he insisted.[22]

But Peale readily acknowledged that he was introduced to the connection between physical health and mental attitude by Christian Science and the other healing sects. When he was in seminary, he claimed, such topics of personal religion were ignored in favor of studies of "ethical insight, moral precepts and sociological improvement."[23] As a consequence, until he learned from "Christian Science and Unity and other such movements" of the spiritual dimension of health, he held the mistaken notion that medicine was "solely materialistic." It was only in later years that he said he had come to appreciate the decisive role played in healing by one's own thoughts and emotions.

There were apparent limits to the mind's ability to control the body, he conceded. People did die. They did get cancer, and heart disease, and their eyesight failed. Yet when considering the question of whether positive thinking always works, Peale answered with an unsurprising "yes." In the course of our lives, he said, it was the dark thoughts—depression, discouragement, hatred—that made us sick, But these disabilities could be instantly overcome if people accepted the miraculous healing power of the Great Physician.[24] The problem was that people had to make a total commitment, to "go all out for it," if healing was to occur, and since so few were willing to do that, they remained sick in body and soul. When the terminally ill did use positive thinking and affirmative prayer, they died "courageously, gallantly, going into eternity with glory."[25] Positive thinking, he claimed, allowed the individual in pain to "rise above the pain until the pain no longer masters him, but he masters it": It does not remove the pain, it recognizes the negative, but keeps it in "proper size."[26] His scientific Christianity, he maintained, provided "workable formulas" for long life, health, vitality, and happiness, but it depended on the quality of one's inner life: the need to be "in harmony," to cultivate an "inner relaxedness," to keep life free of hatred and resentment. Personal religion, practical Christianity, taught one how to do this.

The ambivalence suggested by this view of healing revealed itself again when Peale treated the ultimate enigma of human existence, that of death. Evangelicals who might be confused about his concept of "inner relaxedness" were familiar with his definition of hell. When he posed the rhetorical question, "What do I know about hell after death?" he answered, "I believe in the Bible." "I can see for myself that there is a law of life by which like attracts like," which means that "if you behave badly you will end up in hell."[27] Yet his preferred vision of death, one to which he returned frequently in sermons, had more in common with the mental healing tradition, including its assumptions about the reality of psychic life, than it did with evangelicalism. He analogized the passage of life to death with the movement from the womb to life: It was simply another stage of existence.

To those who claimed that the power of unconscious mind was a form of "clairvoyance," he explained it as "spiritual telepathy" accomplished by "the passage of God's thoughts through the mind." It was rather like Emerson's transparent eyeball. Peale studied the literature of psychic research and was familiar with J. B. Rhine's experiments at Duke University. With neither embarrassment nor hesitation, he revealed in sermons and speeches his personal experiences with psychic phenomena. Most of his references were to encounters with intimates who had died, such as the time he felt the touch of his mother's hand on his head after her death, or when he saw his father's vigorous image approaching him on the platform at Marble Church, or when he recognized his deceased brother Bob smiling at him and greeting him through a brick wall at the Foundation for Christian Living. And there were many others to which he bore witness, at least one of which was picked up and headlined by the sensational *National Enquirer*. Within his Easter sermons orthodox and alternate perceptions of death often competed, with a traditional view of the resurrection sharing place with a metaphysical understanding of death as movement to another level of existence.

Peale's interest in questions of health derived primarily from his view of personal religion, but he was certainly sensitive to the fact that it was the chief expressed concern of his supporters. According to tallies taken by the various agencies within his national ministry network, health matters were the primary concern of Peale's constituents.

The most systematic record of audience priorities was kept by the staff of the Foundation for Christian Living. The mechanism for collecting this data was an annual Good Friday prayer request service. At its inception the FCL staff met daily to pray for persons who wrote or called in for help, a practice they called "absent affirmation" or "absent healing," as in the tradition of "Silent Unity" employed by the Unity group. After 1970 these daily sessions became weekly ones, with Good Friday set aside as a special occasion to respond to these requests. The sample letters became an in-house guide to the needs people identified, an important measure of both constituent concern and a growing trend within the culture toward non-orthodox forms of healing.

A staff analysis of requests received between 1971 and 1986 confirmed that health was consistently the most important concern to those who wrote.[28] Evidence showed that while a single majority of letters, that is, close to one third, asked to have prayers said for particular individuals named in the request, almost an equal number asked for prayers specifically for health, and often the two categories overlapped.[29] In deeply personal letters, writers identified themselves, not only by name and address but by marital status, family size, interests, and occupation, pouring out intimate details of their psychological and sexual lives, perhaps accomplishing just in the writing—as some FCL staffers suspected—the therapeutic relief they sought. A clear majority of the writers were married women—a safe estimate is about 75 percent—who spoke of general debilities, headaches, rashes, insomnia, stomach pains, lameness, and fatigue, much as women in an earlier generation had done and were diagnosed as neurasthenics. But these latter-day letter writers also noted middle-class life-style illnesses: anxiety and stress related to a family member's abuse of alcohol or drugs, a husband's job insecurity, an adolescent's self-destructive behavior, social isolation attendant on a move to a new town or neighborhood. They wrote, too, about somatic illnesses, such as cancer, and about the consequences of the disease more for friends and family members than for themselves. These were the maladies that in the early years of the century Elwood Worcester treated at his Emmanuel healing clinic as "functional" illnesses and that specialists at the end of this century would identify as behavioral illnesses. In Peale's case, as in Worcester's, many of his subscribers were seekers, people who had tried other, usually conventional, means for gaining health and had been unsuccessful. Many seemed to find in their interactions with the Peale network a personal touch absent from the scientific impersonality of the medical community.

Peale's message of personal religion during these years, notably between 1960 and 1980, produced moderate, steady organizational growth. Statistical evidence, particularly for *Guideposts* and the Foundation for Christian Living, showed a consistent, cyclic pattern of growth—that is, growth followed by stability and then new growth. This pattern contrasted with the experience of mainline Protestant churches in which defections and membership loss were the new norms. Significantly, Peale's own church, the mainline Marble Collegiate, was not exempt from this general pattern of attrition. While his personal, public ministry flourished, Peale struggled to keep the membership of his home parish stable. In part this was related to Marble's rock-solid base in tradition, which made substantial change difficult. And in part it was due to the way Peale allocated his energies.

Membership records belied the nature of the problem at Marble Church: The church had 3,500 members in 1967 and 3,500 in 1987, seemingly holding its own as an urban congregation as white, middle-class Protestants moved in droves to the suburbs.[30] To maintain its membership at that level, however, required diligent effort by the ministerial staff, for the congregation was a transient one, with people leaving for reasons of employment, relocation, or dissatisfaction almost as fast as new ones joined. According to the Reverend Arthur Caliandro, Peale's successor, the membership was largely regional, its participants coming from within a sixty-mile radius, with 50 percent from Manhattan itself.[31] They were transplanted New Yorkers, and a majority of them, about 60 percent, were women. Even as Peale concentrated his efforts on his extramural ministry the tourists continued to pour into Marble to hear his Sunday sermons, but most did not stay long enough to help pay the bills. The magic he worked in his national ministry was not as effective at home. But it was also the case that his expanding national ministry was commanding more of his attention than the local ministry he began in 1932.

In his approach to holistic medicine, Peale anticipated the burgeoning interest in self-care, medical self-help, and forms of behavioral medicine that emerged in the 1980s. His message was certainly relevant to the Twelve Step program for personal health. It is revealing that, while he was never fully accepted by the leadership of mainline religion, Peale received a more positive, occasionally enthusiastic, response from the medical community.

A Missionary to American Business

A more powerful and controversial emphasis in Peale's ministry in the post-1960 years was personal achievement. Like his message of self-healing, it corresponded to movements in the culture as economic expansion and increased higher educational opportunities augured new possibilities for middle-class Americans. The emphasis was also consistent with his redesigned ministry to business groups, with whom achievement was often, though not always, a euphemism for financial success. His love affair with the business world was mutual. When he finally retired in 1984 and was completely free to schedule his time himself, he elected to spend the great majority of his public time—he said "all" of it—in the realms of commerce and finance. The appreciative audience revealed its regard for him over the years by awarding him hundreds of plaques, which lined the walls of the corridors of his new and sparkling Peale Center for Positive Thinking.[32]

Even earlier, Peale's critics had disparagingly identified him as pastor to the entrepreneurial class, a pacificator of their moral doubts and a defender of their image of laissez-faire. As a portrait of his ministry, it is thin and incomplete. While he did court the company of the barons of business, it was often with an eye toward gaining a contribution for one of his projects, a practice many evangelists before him had found vital to their work.

But it was also true that Peale admired businessmen, enjoyed their company, and found a personal identification with salesmen in particular. In his lexicon of accolades, he reserved his highest tribute for the person he called a "man's man," or one of "the real boys," and they were often men of industry. He once observed that "Jesus Christ always gets real men," that "He doesn't go after the willy-nillys," although he might have suspected that some of his own critics in the seminaries and universities were of the willy-nilly type. When he reflected on his career in his old age, he decided that if he had not gone into the ministry he would have enjoyed the business world. Surely his wife, he noted correctly, would have been an excellent business executive, as indeed she was at the Foundation for Christian Living.[33]

Salesmen, the butt of jokes and scorn and caricatured like the tax collector in biblical times, were his particular favorites: "If I were not a minister," he once said, "I think I would be a salesman."[34] He wore the title of "God's Salesman" proudly. He liked successful salesmen because they were "dynamic," engaged in a "great work in life," trying to persuade people to "walk a road of agreement" with them so that as customers they might be supplied with goods and services that would benefit them. Reversing Arthur Miller's stereotypic failed figure of Willy Loman helped account for Peale's enthusiastic welcome at sales meetings and motivational training sessions, but there was also an obviously high level of shared regard. Peale's personality opened and blossomed in unique fashion at sales conventions, and his audiences were effusive in their response. Salesmen—in Peale's later life, salespeople—were, like himself, often tied to nomadic, endlessly mobile existences, aware that every transaction was an individual performance and a personal challenge. But more than any other occupational category, salesmen exemplified the new and fragile nature of achievement-based success in modern society: With their every sales encounter a crisis of sorts, they were the ideal model for Peale's message of practical religion.

His self-appointed ministry to America's businessmen actually began relatively early in his New York tenure. It was in the late 1930s when he said he noted the disparity between the predominantly female congregations he preached to every Sunday and the exclusively male audiences he met during his weekday speaking engagements, and he made his goal a ministry to businessmen, either in the church or in their workplace. According to his son John, recalling family history for a reporter's interview in 1976, "a big change came at the end of the thirties when his father stopped being theological and ministerial and became interested in specializing to people in the business world."[35] In the early years, train travel set the boundaries for his range of effectiveness, but once Peale overcame his fear of flying he pushed himself and technology to the limits. As his work in the business community expanded, presumably the generous honoraria he received for his appearances helped fund new projects in his national ministry.

But in addition to the emotional and financial rewards Peale enjoyed from his work as "God's Salesman," he also knew that there was a religious justification for it in the tradition of New Thought and mental science, particularly in its emphasis on the "law of supply." One of the affirmations of the International New Thought Alliance, the concept of divine supply grew out of the belief that a beneficent creator God would not deprive humanity of what it needed for living full, abundant lives. One of Peale's favorite biblical texts, "I am come that they might have life, and have it more abundantly" clearly held interpretive potential in terms of this sense of supply. The "spiritual law of abundance" was set in motion, he explained, at the moment a person admitted Jesus into his or her life, transforming an "ineffective individual" into a "triumphant personality."[36] Even then, however, the law of abundance would be blocked if an individual failed to stay "in tune with God." The transformation from failure to success required energizing one's inner strength.

Although Peale often argued that there was "nothing wrong with success," that God had promised abundance and that to wish the realizable wish was not a sin, he recognized that his seeming support for a success ethic made him more vulnerable to charges from liberal critics. To counter such objections he contended that the successful life not only began with the acceptance of Jesus but required as well the forgetting of self and personal gain. Though it sometimes appeared that it was the wicked who prospered, in the long run, and by the so-called "law of averages," Peale said, things averaged out, so that "those who insisted on living a bad life, ended up bad."[37]

More generally, however, he agreed that most people wanted "adequate material blessings," as he did himself, and he assured his congregation that it was unnecessary "for anybody to live apart from God's law of supply."[38] But it was helpful to stay mindful of the significance of Jesus' parable about the disciples who came up empty when they cast their nets on the left side of the boat: When you fish on the left side, you shut off the law of supply and catch only failure, but when you change to God's side your attitudes change and then everything else changes.[39] Therefore, whether the subject was fish or success or health, the principle was the same: One needed to stay in tune with the Infinite to achieve one's hopes.

Abundance was thus as much an aspect of the realizable wish as good health. The techniques Peale advocated for dealing with illness and pain—picturize, prayerize, actualize—were the same ones he applied to material rewards. He read extensively in success literature and frequently invoked the success writers Napoleon Hill and W. Clement Stone, reminding his audiences that "what the mind can conceive and believe, the mind can achieve."[40] The "Christ-motivated mind" presumably held limitless power, including the power to help individuals achieve personal goals in the material world.

Achievement was not confined to financial success, but as interpreted by Peale was intended to apply to the whole range of human efforts to reach personal goals. And it was received that way by many; it was especially attractive to those in performance-based competitive situations, from opera singers and salespeople to competitive swimmers and track stars—but it suggested possibility to anyone with ambitious personal goals. Peale considered his critics as those "super-pious people" who have forgotten that a "good, wholesome desire to make something of yourself . . . still beats in the American breast."[41] By combining patriotism with achievement and the success ethic, he cast a wide net and won the attention of an extraordinarily diverse segment of American culture. He located achievement-oriented supporters not only among evangelicals and New Age novices but even among some readers of the interracial publication *The New Day*, begun by Father Divine.[42]

Significantly, when the Sermon Publications project of the Foundation for

Christian Living changed its format in 1984, it adopted *Plus* as its new name, a reminder that it was about the "plus factor" in life. Along with Robert Schuller in California, the *Plus* editors pointed out that the plus symbol was also the sign of the cross. The theme of the publication, however, was more consistent with William James's concept of the "more" life of "going beyond."

The best measure of Peale's affection for those in the world of commerce and industry was his never-ending desire to be among them, to accept their invitations to speak and receive their awards. Into his nineties, he remained an itinerant, continuing to log 200,000 miles annually, collecting accolades, awards, and honoraria as he commuted from one sales convention to another. Although he idealized home and family, his personal symbol was rather a flight bag than a latch key. On rare occasions in retirement he extracted moments for undiluted pleasure, as he did for a few weeks in 1987 when he and Ruth took their whole family, grown children and related dependents, on an African safari. But more typically, Ruth was his only traveling companion, and their travel was work related. Although they regularly ventured overseas in the interests of the Foundation for Christian Living, visiting places as diverse as Korea and South Africa, their more usual commitments were to engagements at home.

In advanced age he became an even more arresting model of achievement. In 1986 when he was eighty-eight, in a week not atypical on his calendar, he addressed four different sales and management meetings at four different locations, ranging from Las Vegas to San Francisco to New Orleans to Houston.[43] His schedule for that year hinted at the catholicity of his interests: There were speeches to the Amway organization, the congregation of Schuller's Crystal Cathedral, Ken Blanchard's Training and Development Conference, ordinary Methodist churches, and Domino Pizza executives, as well as scores of other groups. In addition, through 1988 he was regularly invited to address the gathering of evangelicals and fundamentalists at Ocean Grove, New Jersey. In recognition of these extraordinary efforts, he received the 1987 "Marketing Communicator of the Year" award. It became redundant for his presenters to cite him as the living testimony to his own message of practical Christianity: At advanced age, he was physically and mentally fit, still creating personal goals.

The Clergy and the Country Club

For Peale, the reverse image of this personally empowering practical Christianity was what he termed "nominal Christianity," or the creedal religion of the institutional church. In his vocabulary, nominal Christianity was a term interchangeable with bureaucracy and establishment, and thus had a large meaning for him as for many in his audience. Where practical Christianity was understood to be dynamic, democratic, vital, and organic, nominal Christianity was identified as cold, formal, bureaucratic, stuffy, and elitist. It was this view of practical Christianity, or popular religion—born of the culture and his schooling and filtered through the lens of childhood memories—that led him to spend his lifetime contesting with those he considered members of a privileged religious bureaucracy.

The social tumult of the sixties gave him new reasons for challenging the values of the liberal establishment. Everywhere in the culture he found evidence of "decadence" and social decay. Foremost among the problems, he believed, was the decline in personal morality, particularly in sexual matters, and he saw its beginnings in the loose social fabric cultivated by liberals. They were the people who for decades, Peale argued, had supported churches that concentrated on preaching about "social programs" instead of "personal morals." It was they, too, he argued, who in the name of "civil liberties" failed to protest the growth of pornographic literature. "With these people," said Peale, "it's all liberty and none of it's civil." They were the uninterested parents who belonged to the "country club" set and who, though church members, worshipped at churches in "a swanky, super-duper neighborhood," where presumably the cycle was regenerated by more preaching about social issues. "In less favored neighborhoods I would expect more morality," he observed.[44]

Since, ironically, Peale believed the clergy had more power than they really did, he found it a simple equation to connect these large problems with the churches. The churches he held responsible he associated with a particular social class and a religion of economic privilege, where he believed such socially questionable issues as a free press for pornography took precedence over pragmatic concerns such as child-rearing.

The solution seemed self-evident. It required looking backward to the values of a less complicated past, but also accommodating to the future needs of a "modernized" church. In studying the past, Peale found a ready analogy to the contemporary scene in the history of the early church. In the earliest days of Christianity, he said, the "simple ministry" of Jesus was the example for "real Christianity," unlike the "nominal" type that prevailed in too many churches of the present. He believed that the first-century Christians emphasized "the dignity and greatness of man under God and . . . the resurrection of the soul," but all that changed when Christianity became "the state religion of the Roman Empire" which was "the worst thing that ever happened . . . to it."[45] Then the "simple message of Galilee" got complicated by "theology, philosophy, and sacerdotalism," and as it grew in favor with the state, was distorted by "ceremonials, . . . parades and processionals and robes and candles and incense."[46] The lesson for the modern age was clear enough: The time had come, said Peale, for the church, as for individuals, to move away from "nice ethical ideas" to religion based on "depth experience."[47] For too long, he contended, the church had been "clinging to our history and to old traditions," but now the needs of the modern era demanded change, and the only viable option was a personal religion of experience.

Peale admitted that the church had tried to minister to the needs of the age, but in most cases its efforts were applied in the wrong direction. They only replicated the errors of the past, making religion more "intellectual," more concerned with "formalism," and more "bureaucratized." His dissatisfaction with the National Council of Churches was part of this larger indictment of the church in general. He persistently chided the Council for its bureaucracy, suggesting on one occasion that a possible outcome of the merger of denominations could be the creation of a "lot of Protestant popes"—a statement that predictably led to another round with his critics.[48]

His argument with the Council effectively summarized his populist agenda: Its program compromised "freedom" and "the right to dissent," granted leadership to the liberal churches that were growing "weaker spiritually," and expected to change society through "super-duper social improvements."[49] He frequently posed the question, "Do you have to lead a strike to be a Christian?" Church leaders had neither convinced him of the need for merger nor of an effective plan for evangelism: Weren't these the very people, he wondered aloud, who had criticized as superficial the religious revival of the fifties which had brought unprecedented numbers into the churches?[50]

When the Reagan administration brought renewed prominence to evangelicals, Peale found the change heartening. The shift looked more like a vast sea change as he took its measure in 1989: "Christianity began as a movement," he said, "then it became an institution, and now it's becoming a movement again."[51] He welcomed the evangelical renewal, including the pantheon of televangelists, and cheered what he thought were small "cells" of believers springing up everywhere, even within business. He felt he knew from personal experience what it meant to participate in a return to "depth religion." His seminary education, he said, had left him a skeptic, with an "intellectualized religion" requiring him to "struggle for years" to recover real faith. Intellectuals generally, he believed, were fearful of acting on faith and were willing to "accept every kind of defeat for society as well as individually."[52] As for himself, he had come to enjoy the simpler expressions of faith preached by the new evangelicals, listening to the television evangelists whenever he had the opportunity at home. On one occasion, moved by listening to an appeal from the Reverend Rex Humbard, he felt part of the audience asked to commit its life to Christ, and alone in his room, he raised his hand along with the rest.[53]

Life after Marble

Eventually, reluctantly, however, Peale recognized that the time had come for him to go, to leave Marble Church to his experienced apprentice and successor, Arthur Caliandro. He delayed retirement well beyond the usual age, in part because local church leaders were convinced the attendance would fall off if he were not there and in part because he found it difficult himself to abandon a role that, like his marriage, had so defined his life. In 1984 he had been at Marble fifty-two years, having arrived at a small, dispirited congregation in 1932 as an exuberant thirty-four-year-old Methodist with a reputation for making churches grow and leaving as an octogenarian with the early promise fulfilled. An existence without Marble was unpredictable, but far easier to contemplate than life without Ruth. At his request, he retired with very little fanfare.

Leaving the church did not mean leaving work, but rather reordering priorities and enjoying the pleasure of being more selective about which invitations to accept. At eighty-six, he had come to realize one of the great goals of populist leaders generally—that is, to be an embodiment of the values and hopes of their constituency. A telling challenge to the calendar, vigorous, healthy, and alert, Peale became the living testimony to the application of practical Christianity. But he had had reminders that the aging process could not be halted indefinitely. Still performing without notes, there was a time when he had groped painfully for an elusive thought without success, until the awful silence was broken with an idea thrown to him by Ruth from the congregation. Still, he appeared a vital witness to the power of personal religion, with only a bit of arthritis in his fingers and some hearing loss the most obvious signs of his birth date in the nineteenth century. When he left the church, he retained his position as Senior Minister of the Collegiate Church System, a seniority post endowed with considerable intramural decision-making power over policy and finances.

Even in retirement Peale maintained a high level of personal goals, a factor that contributed to his sustained celebrity status. He made frequent appearances on television and more frequently was heard on radio. He presided over the reorganization of the administrative structure of the renamed Peale Center for Christian Living, presumably preparing for a smooth transition when he died, with his son-in-law, John Allen, serving as president of the board of trustees and his daughter, Elizabeth Peale Allen, as Ruth's probable replacement. He continued to write and lecture, producing over a dozen books in the next seven years. Like an old-fashioned Methodist itinerant, Peale was constantly on the road, taking his work with him, as between appointments or on flights he and Ruth answered mail, worked on another book manuscript, or prepared for an upcoming meeting. On a day well into retirement, when she was feeling particularly weary, Ruth asked her husband of many years a question whose answer she already knew from long experience. She asked him if he thought they would ever retire, knowing full well that his answer would be "no."[54] Whatever the occasion, the Peales were inseparable, their relationship and shared work the mainstays of their lives.

Their now-combined careers made a fitting final chapter to Peale's life work. He had always maintained that he was only a part of a team ministry, and the evidence bore him out. The only difference in the years after Marble was that both members of the team were, if not equally, almost equally, obvious. Ruth had been Norman's emotional support and the chief catalyst for his ideas, accepting a substantial, if less public, share of the work load. His very ministry of positive thinking would likely have taken a different turn if she had not been available to him as an antidote to his dark and worrisome side. Obviously endowed from childhood with a healthy measure of confidence and inner strength, she provided her family with its sense of stability and was the real possessor of positive thinking in their household. In later life the Peale children remembered their father as a "cyclone tearing the sky" and their mother as its calm center.

Once he left Marble's pulpit, Peale did not return except for unusual events that approached state occasions. Although the Peales kept their New York apartment, they effectively transferred their work and living to Pawling. Their base of operation was the Foundation for Christian Living, where they functioned from connecting offices, she as the organization's competent CEO and he as prolific writer and public personality. Neat and natty behind his desk, Peale kept in touch with the business around him through the aid of two hearing aids, content to let his wife guide the spiritual empire that the Foundation for Christian Living had become.

In a life that spanned nearly a century, Peale had seen important changes in the culture and had willingly offered commentary on many of them.[55] And yet he had difficulty absorbing the most important, personally relevant change to occur during that time—the transformation of his version of practical Christianity from a marginal mental science perspective to a mainstream position. He understood his own publicity well enough, but essentially an old Methodist at heart, he resisted the notion that the downtown mainstream church that he once knew had changed, likely now the home of a lively Pentecostal congregation or a sect of spiritual seekers. He was less than satisfied with the final act of the script he had written.

By almost any measure Peale's ministry had been exceptional. He had built a national and international network of supporters, held together by an innovative communications network. He had enjoyed personal celebrity and financial success. More important, he had seen during his own ministry his emphasis on personal religion take root in fertile cultural soil, his themes of health, achievement, and personal empowerment flowering in a receptive environment. Their success was in no small measure due to the image Peale himself conveyed of the efficacy of personal, populist religion.

During his long ministry, his folksy appeal, starched white shirt, lack of Roman collar, and image as populist preacher made him appear, to paraphrase Grant Wacker, part Norman Rockwell and part McGuffey Reader.[56] In its own way, the image Peale conveyed represented a rebuke to the privileges, style, and intellectual orientation of the mainstream religious hierarchy. More substantially, his preaching and his example were successful precisely because they were attuned to important shifts in the culture just beginning to break the surface—the "restructuring" process studied by Robert Wuthnow that began in the Peale decade of the fifties. Peale's message of personal religion, his striking example of personal autonomy, his expression of feelings that were antiestablishment and hostile to bureaucracy, resonated with significant numbers of evangelical women and entrepreneurial men who perceived a form of empowerment in his gospel of practical Christianity. His was a message resonant with the culture, popular in its appeal, and historic, even ancient, in its invocation of harmonial nature as a source of divine power.

Lunching at a restaurant in Pawling on a beautiful summer day in July of his ninety-first year, the sun streaming in behind him through hanging green plants, Peale speculated about the changes he perceived occurring within American religion, particularly Christianity. It was with obvious pleasure that he noted great shifts within Christianity, from a small movement to an expansive institutional life, until now in its present incarnation it was once again becoming a movement. That was exactly what his own network of practical Christianity had become. It was a national movement, and a kind of shadow church in which many people held a dual form of religious citizenship. By most measures, the movement Peale initiated was extraordinarily successful. The implications of that success were mind-boggling, as well as discomforting, to the elderly son of a Methodist parsonage upbringing: How much credit could he take for the fact that in 1984 a Gallup poll revealed that as many people were involved in positive thinking seminars as belonged to the Methodist Church? A gleam in his eyes, his finger striking the table, Peale affirmed once again, "I belong to this."

Notes

1. Interview, Norman Vincent Peale, Pawling, New York, July 5, 1989.

2. Episcopal activist layman William Stringfellow protested Peale's candidacy on the grounds that he had been "consistently silent" on the race issue and had appeased "the base complacencies of white Anglo-Saxon Protestants." Peale offered to withdraw but was persuaded to stay. He provided the Council with responsible, if moderate, leadership, issuing statements against civil rights abuses and accompanying the Council's paid executive, the Reverend Dan M. Potter, to Washington to press for voter registration. He also expressed a position in favor of abortion. At a preliminary New York meeting to plan a Washington civil rights gathering, Peale ran into John Bennett, who asked if Peale was going to the Washington meeting. Peale said later he assumed it was the first time they were on the same side of a struggle. *The New York Post*, March 7, 1965; *New York Herald Tribune*, March 7, 1965; interview with NVP, Pawling, New York, February 25, 1983.

3. To a woman in Canada, he wrote that a recent attack on the National Council

of Churches was the result of "very vicious, wholly unwarranted . . . false propaganda." NVP to Doris A. Moore, June 23, 1954; NVP Ms. Coll., Box 348. This was a rather typical response to strident criticisms from anonymous laypeople. To businessmen whom he knew, who voiced somewhat similar sentiments, he could be more equivocal, although even to them he denied that a charge of communism could be applied to the Council leadership.

4. NVP, "Christianity in This Age," *Sermons*, Vol. 13–14, 1962–63, Foundation for Christian Living, Pawling, New York. He said, "I have been accused of belonging to both branches [fundamentalists and modernists], and that is a fact, I do."

5. Wade Clark Roof and William McKinney, *American Mainline Religion: Its Changing Shape and Future* (New Brunswick; NJ: Rutgers University Press, 1987), p. 36.

6. Robert Wuthnow, "Religious Movements and Counter-Movements in North America," in *New Religious Movements and Rapid Social Change*, James A. Beckford, ed. (UNESCO: Sage Publications, 1986), p. 1. James Gordon Melton's *Encyclopedia of American Religions* describes 111 new religions that appeared between 1970 and 1977, most originating in 1970 or 1971, compared with a total of 184 for the 1960s. See Wuthnow, "Religious Movements," p. 16.

7. Robert Wuthnow, *The Restructuring of American Religion* (Princeton, NJ: Princeton University Press, 1988), p. 56. Wuthnow adds that "Against cold intellectualism, popular sermons advocated a religion capable of expressing deep inner emotions; against an outmoded social gospel, a message of personal redemption, against ineffectual concern for social ills, the need to care for individual souls" (p. 57).

8. See, for example, Roof and McKinney, *American Mainline Religion*, p. 48. "Beginning with the student movement of the sixties, the quest for self-fulfillment and a more expressive individualism spread into much of middle-class America, especially among managers, professionals, and frustrated housewives. By the late seventies, this culture had spread into more diverse, more alienated sectors of the society, including some minority and blue-collar constituencies." Wuthnow's *The Restructuring of American Religion* has revealed the great diversity of forces at work creating the restructuring of religion.

9. Nathan Hatch, *The Democratization of American Christianity* (New Haven, CT: Yale University Press, 1989), p. 22. Hatch expanded his definition of religious populism as it existed in the New Republic in the following way: "In at least three respects the popular religious movements of the early republic articulated a profoundly democratic spirit. First, they denied the age-old distinction that set the clergy apart as a separate order of men, and they refused to defer to learned theologians and traditional orthodoxies. All were democratic or populist in the way they instinctively associated virtue with ordinary people rather than with elites, exalted the vernacular in word and song as the hallowed channel for communicating with and about God, and freely turned over the reigns of power" (p. 10).

10. Allan R. Broadhurst, *He Speaks the Word of God: A Study of the Sermons of Norman Vincent Peale* (Englewood Cliffs, NJ: Prentice-Hall, 1963), p. 83. Broadhurst analyzed a twenty-year cycle of Peale sermons and identified as the unique quality the recurrent "Thought Theme," or "the message that thoughts and mind attitudes determine the situation of a person's life."

11. NVP, "Technique for Meeting Crisis Days," *Sermons*, Vol. 1–2, 1950–51, FCL, Pawling, New York.

12. NVP, New Brunswick Lecture Notes, 1942, 1944; NVP Ms. Coll., Box 202.

13. NVP, "Your Future in This Chaotic World," *Sermons*, Vol. 17–18, 1966–67; and "Christianity Still Has Life-Changing Power," Vol. 13–14, 1962–63.

14. NVP, "Will Yourself a Better Tomorrow," *Sermons*, Vol. 15–16, 1964–65.

15. NVP, "Why Worry When You Can Pray," *Sermons*, Vol. 19–20, 1968–69.

16. Catherine L. Albanese, *Nature Religion in America: From the Algonkian Indians to the New Age* (Chicago: University of Chicago Press, 1990), pp. 126, 134.

17. NVP, "What Does Christianity Really Teach?" *Sermons*, Vol. 13–14, 1962–63.

18. NVP, "Your Thoughts Can Generate Creative Power," *Sermons*, Vol. 1–2, 1950–51.

19. NVP, "A Secret of Radiant Health," *Sermons*, Vol. 1–2, 1950–51.

20. Ibid. Peale referred to her as Dr. Florence Dunbar.

21. NVP, "Health and Happiness through Relaxed Attitudes," *Sermons*, Vol. 15–16, 1964–65.

22. Interview with NVP, Pawling, New York, June 11, 1987.

23. NVP, "Spiritual Aids to Health," *Sermons*, Vol. 7–8, 1956–57.

24. NVP, "Good Health through Constructive Thinking," *Sermons*, Vol. 11–12, 1960–61.

25. NVP, "Does the Power of Positive Thinking Always Work?" *Sermons*, Vol. 11–12, 1960–61.

26. Ibid.

27. NVP, "Getting What's Coming to You," *Sermons*, Vol. 13–14, 1962–63.

28. Typescripts obtained from the personal files of Kenneth D. Winslow of the Foundation for Christian Living. Annual "Good Friday Prayer Request Surveys," prepared by Winslow, and by Winslow and Mary Lu Verrier, categorize requests by subject. Tables of "24-Hour Prayer Partnership" also show numbers graphically. Foundation for Christian Living, Pawling, New York.

29. A table analyzing "24-Hour Prayer Partnership" revealed the following information:

	1971	1972	1973	1974	1975	1976	1977
no. request slips	7,848	9,838	24,379	8,357	19,468	23,551	21,613
survey of 1,000 prayer slips:							
Health	346	372	377	385	363	559	477
Faith	257	297	251	186	163	249	116
Family problems	110	151	143	172	147	312	120

From 1979, the total for these prayer requests remained about constant at approximately 20,000. According to analyses of these requests between 1979 and 1986, done by Kenneth Winslow and Mary Lu Verrier, about one third (33 percent in 1985, 29.1 percent in 1986) contained names of individuals for whom prayers were asked; close to another third (19.51 percent in 1985, 27.2 percent in 1986) related to issues of health; and questions of faith came next (17.76 percent in 1985, 19.3 percent in 1986).

30. Interview with the Reverend Arthur Caliandro, Marble Collegiate Church, New York City, June 19, 1987.

31. Ibid.

32. Interview with NVP, July 5, 1989, Pawling, New York.
33. Ibid.
34. NVP, "The Greatest Year of Your Life," *Sermons*, Vol. 11–12, 1960–61.
35. *Chatham Star-Tribune*, April 15, 1976, p. 11.
36. NVP, "Let the Law of Abundance Work for You," *Sermons*, Vol. 5–6, 1954–55.
37. Ibid.
38. NVP, "You Can Have What You Want from Life," *Sermons*, Vol. 7–8, 1956–57.
39. NVP, "Changed Attitudes Change Everything," *Sermons*, Vol. 5–6, 1954–55.
40. NVP, "Believe You Can and You Can," *Sermons*, Vol. 19–20, 1968–69. Stone was the kind of self-made businessman Peale admired. He was also a generous contributor to some of Peale's projects.
41. NVP, "The Three Great Secrets of Success," *Sermons*, Vol. 1–2, 1950–51.
42. There are many areas of similarity between Peale's program of positive thinking and Father Divine's Peace Mission Movement. Peale's picture along with a brief column by him appear as an ad in *The New Day*, June 15, 1991.
43. Interview, NVP, Pawling, New York, June 10, 1986.
44. NVP, "The Tragic Crisis in Morals," *Sermons*, Vol. 15–16, 1964–65.
45. NVP, "What Does Christianity Really Teach," *Sermons*, Vol. 13–14, 1962–63.
46. Ibid.
47. NVP, "Christianity Still Has Life-Changing Power," Vol. 13–14, 1962–63.
48. Interview with NVP, Pawling, New York, February 24, 1983. He said the statement about "Protestant popes" had been teased out of him by reporters.
49. NVP, "Health and Prosperity May Be Yours," *Sermons*, Vol. 13–14, 1962–63.
50. NVP, "Christianity in This Age," *Sermons*, Vol. 13–14, 1962–63.
51. Interview with NVP, Pawling, New York, July 5, 1989.
52. NVP, "The Greatest Year of Your Life," *Sermons*, Vol. 11–12, 1960–61.
53. Interview, NVP, June 11, 1987, Pawling, New York.
54. Conversation with Ruth and Norman Peale, Pawling, New York, July 5, 1989.
55. NVP, *The Power of Positive Living* (New York: Doubleday, 1990). The book is an account of his historical memories.
56. See Grant Wacker, "Searching for Norman Rockwell: Popular Evangelicalism in Contemporary America," in Leonard I. Sweet (ed.), *The Evangelical Tradition in America* (Macon, GA.: Mercer University Press, 1984).

Epilogue

To consider the future of Pealeism beyond the lifetime of its creator pushes the boundaries of historical inquiry. Peale ensured that the agency most closely tied to his identity, the Foundation for Christian Living—or more appropriately the Peale Center for Christian Living—was prepared to continue the line of succession with his son-in-law and daughter in charge. Without the constant replenishment of its literary tradition by Peale himself their work was cut out for them. The national network of people and preacher that became the Phenomenon of Pealeism was held together by Peale's image, surely, but more important, it was sustained by the appeal of his unique message to a substantial segment of mass culture.

A useful guide to the future of Pealeism is the historical experience of other religious populist movements in American life, a tradition that reaches back to the seventeenth century at least. Populist insurgencies have usually arisen at moments in time when the culture is destablized and a sense of crisis is prevalent. Populism generally is a byproduct of organized society, and especially prolific in democratic environments. Nathan Hatch, in fact, described religious populism as an important agent in the democratization process in America, observing that "Religious populism, reflecting the passions of ordinary people and the charisma of democratic movement-builders, remains among the oldest and deepest impulses in American life."[1] But the appearance of populist movements raises cultural hackles and consequently calls for its suppression, troubling as it is to those on the right when it appears as a radical movement, and to those on the left when it suggests reaction. This controversial past of religious populism has extended into the area of historical record-keeping, complicating the task of assessing the life history of populist eruptions and their subsequent impact on the culture.

Nathan Hatch's perceptive analysis of religious populism—that it grows out of a grass-roots democratic base and finds its appeal among common people rather than with uncommon individuals—simultaneously describes the nature of the phenomenon and the difficulty involved in evaluating its objectively.[2] The partisanship of the disputants tends to carry over into the scholarship about the movement, with the historians of the events often just those uncommon individuals antagonistic to the anti-intellectualism inherent in populism. The scholarship of those in other fields—notably anthropology and religious studies—however, has introduced analytical perspectives that make it possible to offer several generalizations about religious populist phenomena in America. In the first place, as antinomian insurgencies they have been time-limited, with a long beginning, a brief period when they crest, and followed by gradual decline, a cycle usually correlated with the life of the group leader. In the second place, they have espoused a dualistic message, one that expresses both a desire to recapitulate the purity of a time past and to adapt that perceived pristine past to a bureaucratically corrupted present. Third, the aspect of the message that reaches deepest into the memory treasure of popular culture, defining the source of populist power to challenge the establishment power structure, has tended to extract a belief in supernatural authority respected by ordinary folk and condemned by elites: Typically, populist leaders have been labeled heretical purveyors of the occult. And fourth, the objectives of a religious populist movement are totalistic: The participants envision a major, "democratic" restructuring of the society.

If Pealeism follows the pattern of earlier examples of religious populism, then the gradual decline in the visibility of the leader and the urgency of the message seem inevitable. The ability of modern technology to preserve the past and make it appear alive has made it possible, and in the short run can continue to make it possible, to keep the movement a part of the contemporary cultural scene. But while the urgency of the message and the movement may abate, it is also true that the concepts have so penetrated the culture that they no longer require an active evangelist to gain an audience. Once a vanguard effort to produce social change, forms of Pealeism, at least for the time being, are part of the mainstream. They exist in evangelical clusters, in New Age centers, in Twelve Step groups, and in psychosomatic healing clinics. Peale's message of positive thinking anticipated the appearance of a groundswell shift in the center of power, away from liberal bureaucracy and toward individual autonomy. It is a perception with a basis in reality in such areas of personal choice as religious affiliation, political parties, and health maintenance, although having less firm grounding in reality in the realm of economic and state power.

The dualistic message of Pealeism coincided with a prevalent social ambivalence about the relevance of cultural values, old and new, and concerns over how well the nation was positioned morally and ethically to face an uncertain future. Peale's life provided a kind of living laboratory. By personal history and training attached to Methodism, he was also alienated from the mainstream tradition of modern Methodism and attracted to the simple, practical, and functional approach of metaphysical religious science. Peter Williams has described this ambivalence as one of the qualities of religious populism. Williams has said that popular religion can best be understood as a series of cultural rings:

> The first . . . is the core of symbols provided for it by a great tradition. Secondly, and at its heart, is the system of elaborations and interpretations which it works on these themes. Third, there exists around the edge . . . a third circle,

> which consists of a broad collection of beliefs and practices which once made sense as part of . . . an earlier primitive religious system, but which now are incompatible with the new synthesis.[3]

According to this definition, religious populism can therefore be regarded as an attempt by a noncore group in the culture to make sense of profound social changes by reinterpreting the symbol system of the "great tradition" in ways that are culturally specific to the group. Since its development may also include borrowing from the older "primitive" religious system, it leaves the movement open to charges of dark, occultic beliefs. Religious populism is therefore inherently dualistic, liable to be pulled toward the core and develop more traditional qualities, or swayed toward the periphery and evolve in ways more "primitive" and closer to folk culture.

Peale's movement can readily be distinguished as an attempt to synthesize these two forces—the "great tradition" of Methodism and the peripheral experience of positive thinking—into a modern socioreligious expression. Significantly, he referred to himself as a "child of Methodism" and as one who walked "the broad middle path" in his life experience. As an adherent of the great tradition, Peale made a substantial and lasting contribution to mainstream religion through his role as a popularizer of modernized, clinically based pastoral counseling. It was another medium by which positive thinking was adapted to, and received by, traditional mainstream culture.

By politicizing his views, however, in an effort to effect broad, totalistic, cultural change, he left a legacy more appealing to the two religious groups most strengthened by postwar developments in society—evangelicals and advocates of New Age beliefs. Evangelicals warmed to his conservative politics, and New Age believers appreciated his mystical appeal as well as his relevance to secular human potential ideas. As Peale's movement developed and aged, it found increasing appeal within these two groups: It was absorbed and homogenized within evangelicalism and it was accepted while remaining distinctive within human potential-New Age organizations. The movement's most lasting presence is apt to be experienced within this latter group.

This potential outcome, that the distinctive, unassimilated aspect of Peale's views could have a more lasting influence on the New Age-human potential movement, lends support to the assumption that the essence of Pealeism was the message rather than the messenger. It is also an irony of history that Peale's memory might endure longer among a group drawn not from the "great tradition," but rather the periphery, since Peale always thought of himself as part of the older tradition. By temperament and experience he was more comfortable with the religious tradition of the mainstream than of the margin.

One of the paradoxes of the messenger of positive thinking was that he was acutely sensitive to the reality of darkness and tragedy, persistently conflicted by the experiences of hope and despair in the human condition. Few of Peale's clouded feelings appeared straightforwardly in his speeches or writings, apart from references to the "inferiority complex" that beset him in youth and remained threatening the rest of his life. Yet they were feelings that undergird his worldview and consequently informed his public message. The self-assured "minister to millions," frequently described in the press as "sunny," had another less bright, less effervescent side, most apparent in his combative political activism. In his public ministry, however, he wore the mantle of the popularizer as he gave voice to the expectations and optimism he believed his constituency anticipated from him. His dark side remained veiled, except from Ruth, the true positive thinker in his household. It is possible that his audience recognized this alternate side to his nature and found him even more engaging because he was more like themselves. Peale may have been able to identify with an audience sense of anxiety and despair not because he denied their reality but because he was acquainted with them all too well.

As the leaven that lifted Norman's frequently sagging confidence, Ruth Peale was more than just the other half of the team; she was the strength that sustained the effort. The relationship that she and Norman had was extraordinary by almost any standards, their unusual compatibility the basis for a highly productive partnership. In a marriage of over sixty years, they remained the most important consideration in each other's lives. Ruth had a prescient sense of their mutuality early in their marriage when on a rare trip alone she wrote him a letter that began "My darling:" "Sometimes I think it is cruel," she wrote, "for two people to love each other as much as we do, and to depend so entirely on one another for happiness. Yet that is the case; there is no happiness without you my sweetheart."[4] They continued to depend on each other over the years, for although they cultivated acquaintances, they had no close friends and no apparent need for any. Those who worked closely with them acknowledged the decisive role Ruth played in all matters. Some wondered if Norman didn't actually have the easier job, taking the credit for his speeches, books, and guest appearances while Ruth did the planning and work behind the scenes.

Ruth's emotional consistency and durability were frequently countered by her husband's gloominess and forebodings of difficulties, even disasters. At one point in his life, in addition to the stocks and bonds he and Ruth held, Norman owned a dozen life insurance policies.[5] His view of life often hinted at a mainstream religious sense of transcendent judgment and borrowed from a shaded, Calvinistic conception of the future: He worried over a Communist takeover in the postwar years, an imminent nuclear attack during the 1960s and 1970s, and the destruction of the social fabric of society almost constantly. He was often inclined to see his personal future and the future of his projects in a similar light. He spoke the truth when he said he preached first for himself.

He came by his dark side naturally. His family of origin, for all its Methodist Arminian acceptance of provenient grace, laid hold on a strong tendency toward apprehension and despair. His younger brother Leonard, who also became a Methodist minister, lived uncomfortably in Norman's shadow all his life, a condition for which the elder brother felt responsible. Always ailing, Leonard, Norman believed, was "obsessed by a fear of death."[6] His older brother Bob, a physician, suffered from personal problems that affected his health and his work. Overwhelmed by difficulties, he once wrote to Norman,

> The underlying unhappiness and misery in the lives of people and our inability to correct the causes, makes me feel useless and deprives me of any real desire to go on trying. Religion is not a cure—it is an analgesic that softens the blows and makes it possible to live like those around us, while we wait for the happy day when God will take us home to a place of real life and happiness.[7]

Nor were Peale's parents important sources for a philosophy of positive thinking. His father likely had an ebullient personality, pleasant and winsome, while his mother confessed to being prone to "quick words." She was also sickly for a number of years and gave evidence of being depressed. In her last illness she wrote to her son, "I have very little hope as to the future but my trust is in God."[8] His father, like his brother Leonard, had a great fear of death, and once told Norman that premonitions about the future brought "great stress and agony of mind."[9] Although generally circumspect with his parents, Norman occasionally revealed doubts of his own. Writing to his father after another failed attempt at dieting, trying to get some of the 189 pounds off his 5′8″ frame, he observed, "Perhaps it would be just as well to keep your weight, eat everything, and suffer and die young as to mess around cutting down your weight, discipline yourself, and then still having the same result."[10]

Peace of mind did not come easily to Peale, and contentment was usually elusive. The personal price he paid for being a popularizer, for living out the image he believed his constituency expected, was very high. An advocate of relaxation, he confessed to hating to go to bed because there was nothing to do. That he had been taught by a grandmother when he was a child to check under the bed at night for any suspicious signs may have further complicated his ability to rest easily. On those nights in adult life when he was unable to sleep, he experimented with a variety of techniques: Once he prayed by name for all the people in his apartment, which left him sleepless but surprisingly "refreshed" to face the morning; another time in his darkened study he repeated a biblical verse in mantra fashion—"they that wait upon the Lord shall renew their strength; they shall mount up with wings as eagles"—which did indeed induce drowsiness.[11] There were other unspoken conflicts. Peale warred against the ambiguity and intellectual subtleties of modernism at the same time that he readily embraced the tools of modern technology. Reared with essentially the same core beliefs as his constituency, he was eventually distanced from them by reasons of education and the demands of public life.

And significantly for his ministry, while he believed he faithfully reflected the values of his audience, he learned during the Kennedy campaign that he had seriously misjudged the priorities of many of them.

His celebrity status further complicated his populist image. The holder of three university degrees, a voracious reader, and a dedicated international traveler, he retained a midwestern accent, which made "narrow" sound like "narra" and canceled final "g's" on many verbs. He was a longtime resident of Manhattan's Upper East Side who dressed in bespoke suits while appearing the small-town banker with a Will Rogers accent.

The consequences of celebrity were more cosmetic than substantial, however; on the core issues that mattered, religiously, sociologically, personally, he and his conservative evangelical audience were in harmony. In his call for personal religion, he was voicing his and their critique of all bureaucracy and its failed promises. He likely spoke for them as well when he criticized the social tumult of the sixties, a criticism that had less to do with the political objectives of the student activists than with their social values and behavior, which seemed to fly in the face of traditional middle-class standards. He personally cherished social civility and the norms of a bourgeois social code, patterns he felt were being undermined by the student culture. His own preference for neatness and order went beyond a clean desk, a starched shirt, and organized briefcase to the need to plump up the pillows when someone left a chair. Consequently, although his version of personal religion was premised on a vision of individual autonomy, it drew on older, traditional standards: He found the language, dress, sexuality, bohemianism, and drug use of young people to be evidence not of free expression or a higher consciousness but of licentiousness and vulgar hedonism. He missed the possibility, indeed reality, that there was a link between his version of positive thinking and countercultural affirmations of personal perceptions of ultimate truth. Between the positive thinking evangelical parents of the 1950s and the activist young people of the 1960s and 1970s there was an unacknowledged bond of claims to transcendent truth. Both movements—Pealeism and the student revolt—contributed substantially to the populist surge in the culture.

Peter Williams' observation that "Each cultural and religious system has a logic of its own" obviously applies to Peale's movement of practical Christianity.[12] Perceived by many outsiders as theologically shallow and politically misguided, if not dangerous, Pealeism was sustained by a loyal body of adherents for whom it represented a logical conception of beneficent Nature, a democratic, indwelling force capable of empowering them to gain access to the opportunities and entitlements available in an abundant society. Peale's message suggested the possibility for broad changes in the culture, not in a root-and-branch revolutionary fashion but in a reformist return to better, purer early times.

Peale and his supporters mistakenly placed all the blame for their discontents on the liberal churches—although their failure over the years to minister to individual needs was one source of the problem—when the sources were more deeply rooted in destabilizing processes at work in the culture. The decline in mainstream denominations, the rise of evangelical and New Age organizations, the appearance of "interest groups," all were symptoms rather than sources, consequences rather than causes of these substantial shifts in modern American priorities. Those who have searched for specific causes for the recent changes in denominational loyalty and church membership find that they are not easily isolated but belong to a complex of circumstances related to the "social context," meaning that "a broad cultural shift has occurred that has hit the churches from the outside, and it has hit the affluent, educated, and individualistic, culture-affirming denominations hardest."[13] Pealeism has been a part of that broad cultural shift, in a movement reminiscent of many other evangelical revivals in the past.

Nevertheless, the disenchantment with older forms of belief and authority has persisted long past Peale's salad days. Evangelical and New Age metaphysical groups continue to expand as interest groups and self-help organizations identified with the human potential movement keep proliferating. Significantly, Peale and representatives of liberal religious interests appear to have concluded a truce. In one sense, his critics were correct: He did challenge traditional views of religion, Protestantism in particular. But by their persistent verbal pummeling of Peale, they revealed more about their own large fears than the true nature of the problem. They were justifiably concerned about what seemed like a death struggle for traditional churches.

Peale was in the populist vanguard of the second half of the twentieth century, one who helped pry open the floodgates to change. Personally, on an ideological level, he was closer to his critics than either side was usually willing to admit. Ideological purity held little interest for him, however, and he flourished as a synthesizer, more concerned with what "worked" in meeting the challenges of daily life. It was an appealing approach to persons feeling overwhelmed by the crises in their lives.

But critics asked then and have continued to ask: Is that religion? Robert Fuller has suggested that the various spiritual, therapeutic, and self-help groups that have arisen in the last several decades have created the need to reconsider how people are initiated into the realm of the religious. The exigencies of modern life—its rootlessness, lack of historicity, its sense of "homelessness"—have altered old forms of entry to the sacred. Fuller contends that for their members many of these new groups relocate the "point of access to the sacred in ways . . . well tuned to the psychological and sociological structures of our day."[14] That, of course, is exactly what Peale said he was doing, preaching "the old fashioned gospel" in the language forms of the modern age, a missionary and an itinerant rather than a priest and an educator. The movement was a reminder of the continuing struggle between popular culture and the forces of elite, or intellectual, culture.

The Phenomenon of Pealeism was part of a grass-roots insurgency, populistic and democratic, that reached flood tide in the third quarter of the twentieth century. As the tide gradually recedes at the end of the century, it will be the task of those who inspect its results to distinguish the detritus from the pearls. Pealeism seems destined to leave evidence of its impact as formed and omnipresent channels in the sand.

Notes

1. Nathan Hatch, *The Democratization of American Christianity* (New Haven, CT: Yale University Press, 1989), p. 5. Hatch's excellent study is one of the many revisionist examinations of popular religion to have appeared in the 1980s. Employing interdisciplinary methods, many of these works recast the interpretations of earlier consensus and new left accounts.

2. Nathan Hatch, "Evangelicalism as a Democratic Movement," in *Evangelicalism and Modern America*, George Marsden, ed. (Grand Rapids, MI: Eerdmans Publishing Company, 1984), p. 80.

3. Peter Williams, *Popular Religion in America: Symbolic Change and the Modernization Process in Historical Perspective* (Chicago: University of Illinois Press, 1989), p. 64.

4. RSP to NVP, May 29, 1933; NVP Ms. Coll.

5. Brief and partial accountings of their financial status—some written on scraps of paper—are scattered throughout the Norman Vincent Peale Manuscript Collection.

6. NVP to Clifford Peale (Norman's father), March 25, 1947; NVP Ms. Coll.

7. Robert C. Peale to Norman and Ruth, Sept. 21, 1954, NVP Ms. Coll. Bob was divorced, often experienced financial difficulties, and had health problems that were rumored to include alcohol use. Norman was generous with his resources in helping family members.

8. Anna Peale to NVP, n.d., probably 1936; NVP Ms. Coll.

9. Clifford Peale to NVP, July 12, 1948; NVP to Dad, Mar. 25, 1947; NVP Ms. Coll.

10. NVP to Dad, Feb. 25, 1947; NVP Ms. Coll.

11. Interview with NVP, Pawling, New York, February 23, 1981.

12. Williams, *Popular Religion in America*, p. 61.

13. Dean R. Hoge and David A. Roozen, "Some Sociological Conclusions about Church Trends," in Dean R. Hoge and David A. Roozen, eds., *Understanding Church Growth and Decline: 1950–1978* (New York: Pilgrim Press, 1979), p. 328.

14. Robert C. Fuller, *Alternate Medicine and American Religious Life* (New York: Oxford University Press, 1989), p. 136.

Bibliography

Primary Sources

Manuscripts

The single most substantial source used for the book has been the Norman Vincent Peale Manuscript Collection. Fortunately, and unexpectedly, Peale gave me sole access to this vast, closed collection in 1981. Over the years I worked on the material, the boxes accumulated, so that when I finished my reading of the collection in 1989 there were more than 850 boxes, with additional material coming in on a regular basis from Peale's office. Peale's papers and those of his wife, Ruth Stafford Peale, have both been housed at Syracuse University, although it is possible that they will be relocated to another repository in the future. The material is richest for the years following Peale's arrival at Marble Church, that is, 1932, with very scanty evidence for the earlier years.

Through about 1980, the material is fairly well indexed and cataloged. Since that time, the abundance of documents from Peale's office has increased faster than the library's ability to catalog it. Some of the material is redundant and uninteresting, but much of it is useful evidence for historians of America's social, religious, and political past. Peale's network of contacts was extensive, with the files containing copies of his letters, and often their replies, to national and international leaders in all walks of life. The thick files of letters from anonymous writers, usually either "request" letters asking for help with particular problems or "success" letters describing recovery, provide important commentary on how Americans have described their maladies over the past fifty or sixty years, and why they thought Peale had the answer. To preserve the anonymity of these writers, I have removed their names from the notes.

It is obvious that the material in the files was edited before it reached the library, although seemingly with a light touch. There were no skeletons in the huge collection closet. Nevertheless, I regarded it as my good fortune to have Peale grant me free access to the material and to allow me to produce my own manuscript without his interference or approval. It is to be hoped that Peale and his family will open the collection in the future to other scholars.

The Ruth Stafford Peale Collection is much smaller by comparison, useful primarily for examining her role in the various organizations in which she participated, such as the National Council of Churches.

The other substantial repository of Peale material is the Foundation for Christian Living, now the Peale Center for Christian Living, in Pawling, New York. The new Center contains a well-organized library, with a large collection of photographs, videos, and tapes of Peale's sermons and talks, as well as collections of printed sermons and newspaper clippings. The Pawling complex became the center for Peale's national ministry in the 1960s and has remained the coordination hub for his work. It was the Center that provided me with the complete collection of Peale's published sermons.

Guideposts made available the complete run of the magazine since its inception. The Peales facilitated my visits to the publishing operation in Carmel, New York, and the editing and writing division in New York City.

In the early stages of my research, I was interested in comparing Peale's work with what I perceived as a similar, earlier, mainstream model in the Emmanuel Movement developed by Elwood Worcester. I spent several months one winter working through Worcester material in the Episcopal Archives in Boston and in the Library of Harvard Medical School. In addition, Worcester's grandson, Carroll Worcester Brewster, generously allowed me to look through several scrapbooks of clippings on his relative and the movement kept by his aunt, Constance Worcester. He also arranged for me an introduction to his aunt, who subsequently granted me several interviews. Additionally, in connection with the Emmanuel Movement, I worked through the records of the Jacoby Club housed in the Massachusetts Historical Society. The Jacoby Club was the immediate predecessor to Alcoholics Anonymous, and its records are an important source for anyone interested in issues relating to temperance, substance abuse, and psychosomatic medicine.

Interviews

Interviews provided the other large source of primary evidence for this study. To someone unused to the techniques, advantages, and pitfalls of "oral history," this became a largely self-taught methodological approach. Peale was a skilled interview subject, I a novice interviewer, with the result that much of the early material repeated information available in a number of printed sources. Over the years, I learned more about winnowing the important from the unimportant in interviews, finding some of the best Peale observations, not surprisingly, offered spontaneously over lunch. Most of the interviews took place in Pawling, a couple in New York City, and several in Madrid when I joined an FCL overseas travel-study program for five days in 1981. Ruth Peale was a participant in almost all interviews. I met the other members of his family during the Madrid program and had the chance to speak on several occasions with Peale's daughter, Elizabeth Peale Allen, a member of the FCL staff.

I spoke with many members of the Peale organization in Pawling, and as I have indicated, with my way eased by introductions from the Peales, I was able to interview leaders at *Guideposts* and the Institutes for Religion and Health. I have been to Marble Collegiate Church on many occasions, on two of which I interviewed Peale's associate and then successor, the Reverend Dr. Arthur Caliandro. At other times, I have simply participated as a member of the congregation. Dr. Caliandro made it possible for me to examine records of the Collegiate Church system to follow the changes at Marble during Peale's ministry.

Another interesting source of information was Dr. John C. Bennett, retired president of Union Theological Seminary, living in retirement in Claremont, California, at the time I interviewed him. Bennett's views were useful for their perspective on the 1950s, the Bennett-Niebuhr association, and the Billy Graham ministry, as well as the work of Peale.

I also spoke with members of University United Methodist Church in Syracuse, who remembered Peale's time with them. The church has some records illuminating the Peale years.

Newspapers and Periodicals

For so public a figure as Peale, popular literature is a basic source of information about his interaction with, and perception by, various segments within the culture. Few periodicals missed the opportunity to write something about him, whether the publication was *The National Enquirer*, *Christian Century*, or *The New Republic.* All of his various ministries—Brooklyn, Syracuse, New York—were followed by the local press, and as he gained prominence in the postwar years, he was analyzed by the national syndicates, major papers, and a wide array of periodicals. His own personal clipping file is fairly complete, as I discovered after tracking down articles and reviews on my own.

Peale's Own Writings

Peale's national ministry was essentially a communications network, held together by various forms of Peale writings. In addition to his published sermons, articles in *Guideposts*, and frequent contributions to magazines like *The Reader's Digest*, Peale produced a steady stream of material that appeared in other periodicals. He also authored or co-authored many books: by 1991, the figure stood at forty-one.

Public and Church Records

Most of the documentary evidence I used for church historical information I located at Union Theological Seminary. Useful for following Peale's ministry, and also for viewing it in the larger context of Protestant membership data generally, were: *Records of the Classis of New York: Minutes of the Collegiate Church*; *Annual Report of Religious Bodies.* From the Syracuse office I used the *Records of the Central New York Conference* of the Methodist Church.

Secondary Sources

Listed below are published works I found particularly helpful in developing my ideas about metaphysical religion, positive thinking, and Peale's ministry.

Ahlstrom, Sydney. *A Religious History of the American People.* New Haven, CT: Yale University Press, 1972.

Albanese, Catherine L. *America: Religions and Religion.* Belmont, CA: Wadsworth Publishing Company, 1981.

Albanese, Catherine. *Nature Religion in America: From the Algonkian Indians to the New Age.* Chicago: University of Chicago Press, 1990.

Ashbrook, William E. *Evangelicalism: The New Neutralism.* Columbus, OH: Calvary Bible Church, 1970.

Beckford, James A., ed. *New Religious Movements and Rapid Social Change.* UNESCO, 1986.

Bennett, David H. *Demagogues in the Depression.* New Brunswick, NJ: Rutgers University Press, 1969.

Berger, Peter, Brigitte Berger, and Hansfried Kellner. *The Homeless Mind: Modernization and Consciousness.* New York: Vintage Books, 1974.

Blake, Nelson M. *History of the United University Methodist Church.* Syracuse, NY: Privately printed, 1970.

Bowne, Borden Parker. *Kant and Spencer: A Critical Exposition.* Port Washington, NY: Kennikat Press, 1967.
Braden, Charles S. *Spirits in Rebellion.* Dallas, TX: Southern Methodist University Press, 1963.
Broadhurst, Allan. *He Speaks the Word of God: A Study of the Sermons of Norman Vincent Peale.* Englewood Cliffs, NJ: Prentice-Hall, 1963.
Butler, Jon. *Awash in a Sea of Faith.* Cambridge, MA: Harvard University Press, 1990.
Carlson, John Roy. *Undercover.* New York: E. P. Dutton, 1943.
Carlson, John Roy. *The Plotters.* New York: E. P. Dutton, 1946.
Carter, Paul A. *The Spiritual Crisis of the Gilded Age.* DeKalb, IL: Northern Illinois Press, 1971.
Cashman, Sean Dennis. *Prohibition.* New York: The Free Press, 1981.
Cassequet-Smirgel, Janine. *The Ego Ideal.* New York: W. W. Norton, 1985.
Central New York Methodist Conference. *Annual Yearbook.* 1928.
Central New York Methodist District Conference. *Yearbook.* 1927, 1928, 1929, 1930, 1931, 1932, 1933.
Chandler, Russell. *Understanding the New Age.* Dallas, TX: Word Publishing, 1988.
Cherrington, Ernest. *The Evolution of Prohibitionism in the United States of America.* Montclair, NJ: Patterson Smith, 1920.
Clecak, Peter. *America's Quest for the Ideal Self: Dissent and Fulfillment in the 60s and 70s.* New York: Oxford University Press, 1983.
Collegiate Church. *Yearbook.* 1933, 1934, 1940.
Cousins, Norman. *Anatomy of an Illness.* New York: W. W. Norton, 1979.
Cousins, Norman. *Head First: The Biology of Hope and Healing Power of the Human Spirit.* New York: Viking Penguin, 1990.
Dresser, Horatio. *History of the New Thought Movement.* New York: Thomas Y. Crowell, 1919.
DuBois, W. E. B. *Souls of Black Folk.* Greenwich, CT: Fawcett Publisher, 1961.
Eckardt, A. Roy. *The Surge of Piety in America.* New York: Association Press, 1958.
Ellenberger, Henri. *Discovery of the Unconscious.* New York: Basic Books, 1981.
Ellwood, Robert S. *Alternative Altars.* Chicago: University of Chicago Press, 1979.
Evans, Robert A. "Recovering the Church's Transforming Middle: Reflections on the Balance between Faithfulness and Effectiveness." In *Understanding Church Growth and Decline: 1950–1978.* Dean R. Hoge and David A. Roozen, eds. New York: Pilgrim Press, 1979.
Evans, Warren Felt. *Mental Medicine: A Treatise on Medical Psychology.* Boston, MA: H. H. Carter, 1886.
Frady, Marshall, *Billy Graham: A Parable of American Righteousness.* Boston, MA: Little, Brown & Company, 1979.
Fuller, Robert C. *Americans and the Unconscious.* New York: Oxford University Press, 1986.
Fuller, Robert C. *Alternate Medicine and American Religious Life.* New York: Oxford University Press, 1989.
Galbreath, Robert. "Explaining Modern Occultism." In *The Occult in America.* Howard Kerr and Charles L. Crow, eds. Chicago: University of Illinois Press, 1986.

Galpin, W. Freeman. *Syracuse University: The Growing Years*. Vol. II. Syracuse, NY: Syracuse University Press, 1960.

Gawain, Shakti. *Living in the Light*. San Rafael, CA: New World Library, 1986.

Gottschalk, Stephen. *The Emergence of Christian Science in American Life*. Berkeley: University of California Press, 1973.

Gordon, Arthur. *One Man's Way: The Story and Message of Norman Vincent Peale, Minister to Millions*. Pawling, NY: Foundation for Christian Living, 1972.

Hale, Nathan. *Freud and the Americans*. New York: Oxford University Press, 1971.

Handy, Robert T. *A Christian America*. New York: Oxford University Press, 1984.

Hatch, Nathan. "Evangelicalism as a Democratic Movement." In *Evangelicalism and Modern America*. George Marsden, ed. Grand Rapids, MI: Eerdmans Publishing Company, 1984.

Hatch, Nathan. *The Democratization of American Christianity*. New Haven, CT: Yale University Press, 1989.

Herberg, Will. *Protestant, Catholic, Jew*. New York: Doubleday & Company, 1955.

Hoeller, Stephan A. *The Gnostic Jung*. Wheaton, IL: The Theosophical Publishing House, 1982.

Hoge, Dean R., and David A. Roozen, eds. *Understanding Church Growth and Decline: 1950–1978*. New York: Pilgrim Press, 1979.

Holifield, Brooks. *A History of Pastoral Care in America: From Salvation to Self-Realization*. Nashville, TN: Abingdon Press, 1983.

Huber, Richard M. *The American Idea of Success*. New York: McGraw-Hill Book Company, 1971.

Hughes, Richard T. *The American Quest for the Primitive Church*. Urbana: University of Illinois Press, 1988.

Hutchison, William R. *The Modernist Impulse in American Protestantism*. New York: Oxford University Press, 1982.

Jacoby, Russell. *Social Amnesia: A Critique of Conformist Psychology from Adler to Laing*. Boston, MA: Beacon Press, 1975.

James, William. *Varieties of Religious Experience*. New York: P. F. Collier, 1961.

Judah, J. Stillson. *The History and Philosophy of Metaphysical Movements in America*. Philadelphia, PA: Westminster Press, 1967.

Kelley, Dean M. *Why Conservative Churches Are Growing: A Study in the Sociology of Religion*. New York: Harper & Row, Publishers, 1972.

Kerr, Howard, and Charles L. Crow, eds. *The Occult in America*. Chicago: University of Illinois Press, 1986.

Lacey, Michael J., ed. *Religion and Twentieth Century American Intellectual Life*. New York: Woodrow Wilson International Center for Scholars and Cambridge University Press, 1989.

Lasch, Christopher. *The Minimal Self: Psychic Survival in Troubled Times*. New York: W. W. Norton, 1984.

Levine, Lawrence. *Highbrow Lowbrow: The Emergence of Cultural Hierarchy in America*. Cambridge, MA: Harvard University Press, 1988.

Marsden, George, ed. *Evangelicalism and Modern America*. Grand Rapids, MI: Eerdmans Publishing Company, 1984.

Marsden, George. *Reforming Fundamentalism: Fuller Seminary and the New Evangelicalism*. Grand Rapids, MI: Eerdmans Publishing Company, 1987.

Marsden, George. *Religion and American Culture.* New York: Harcourt Brace Jovanovich, 1990.

Martin, William. *A Prophet with Honor: The Billy Graham Story.* New York: William Morrow and Company, 1991.

Marty, Martin. *The New Shape of American Religion.* New York: Harper & Row, Publishers, 1958.

Marty, Martin, John G. Deedy, Jr., David Wolf Silverman, and Robert Lekachman. *The Religious Press in America.* New York: Holt, Rinehart and Winston, 1963.

May, M. A., in collaboration with W. A. Brown et al. *The Profession of the Ministry, Its Status and Problems.* Vol. II of *The Education of American Ministers* (New York: Institute of Social and Religious Research, 1934).

McGavran, Donald. *Understanding Church Growth.* Grand Rapids, MI: Eerdmans Publishing Company, 1970.

McLoughlin, William G. *Modern Revivalism: Charles Grandison Finney to Billy Graham.* New York: Ronald Press, 1959.

McLoughlin, William G. *Revivals, Awakenings, and Reform.* Chicago: University of Chicago Press, 1978.

Melton, J. Gordon. *Encyclopedia of American Religions.* 2 vols. Wilmington, DE: McGrath Publishers, 1978.

Meyer, Donald. *The Positive Thinkers.* New York: Pantheon Books, 1980.

Miller, Robert Moats. *American Protestantism and Social Issues.* Chapel Hill: University of North Carolina Press, 1958.

Miller, Robert Moats. *How Shall They Hear without a Preacher: The Life of Ernest Fremont Tittle.* Chapel Hill: University of North Carolina Press, 1971.

Miller, Robert Moats. *Harry Emerson Fosdick.* New York: Oxford University Press, 1985.

Moore, R. Laurence. "Mormonism, Christian Science, and Spiritualism." In *The Occult in America.* Howard Kerr and Charles L. Crow, eds. Chicago: University of Illiniois Press, 1986.

Moore, R. Laurence. *Religious Outsiders and the Making of Americans.* New York: Oxford University Press, 1986.

Muelder, Walter G. *The Ethical Edge of Christian Theology.* New York: Edward Mellen Press, 1983.

Peel, Robert. *Mary Baker Eddy: The Years of Discovery.* 3 vols. New York: Holt, Rinehart and Winston, 1966.

Pickering, Ernest. *Should Fundamentalists Support the Billy Graham Crusade?* Chicago, n.p., 1958.

Piper, Watty. *The Little Engine That Could.* Sixtieth Anniversary Edition. New York: Platt Munk Publishers, 1990.

Plymouth Church of the Pilgrims. *A Church in History.* Brooklyn, NY: Stefan Salter, 1949.

Powell, Robert. "Healing and Wholeness: Helen Flanders Dunbar and an Extra-Medical Origin of the American Psychosomatic Movement, 1906–1936." Ph.D dissertation, Duke University, 1974.

Quebedeaux, Richard. *By What Authority: The Rise of Personality Cults in American Christianity.* San Francisco, CA: Harper & Row, Publishers, 1982.

Rieff, Philip. *The Triumph of the Therapeutic.* New York: Harper & Row, Publishers, 1966.

Roof, Wade Clark, and William McKinney. *American Mainline Religion: Its Changing Shape and Future.* New Brunswick, NJ: Rutgers University Press, 1987.

Roozen, David A., and Jackson W. Carroll. "Recent Trends in Church Membership and Participation: An Introduction." In *Understanding Church Growth and Decline: 1950–1978.* Dean R. Hoge and David A. Roozen, eds. New York: Pilgrim Press, 1979.

Siegel, Bernie. *Love, Medicine, and Miracles: Lessons Learned about Self-Healing from a Surgeon's Experience with Exceptional Patients.* New York: HarperCollins Publishers, 1990.

Smylie, James H. "Church Growth and Decline in Historical Perspective." In *Understanding Church Growth and Decline: 1950–1978.* Dean R. Hoge and David A. Roozen, eds. New York: Pilgrim Press, 1979.

Susman, Warren I., ed. *Culture as History: The Transformation of American Society in the Twentieth Century.* New York: Pantheon Books, 1973.

Sweet, Leonard I., ed. *The Evangelical Tradition in America.* Macon, GA: Mercer University Press, 1984.

Tiryakian, Edward A. *On the Margin of the Visible: Sociology, the Esoteric, and the Occult.* New York: John Wiley & Sons, 1974.

Trine, Ralph Waldo. *In Tune with the Infinite.* New York: Thomas Y. Crowell, 1897.

Wacker, Grant. "Searching for Norman Rockwell: Popular Evangelicalism in Contemporary America." In *The Evangelical Tradition in America.* Leonard I. Sweet, ed. Macon, GA: Mercer University Press, 1984.

Weber, Timothy P. *Living in the Shadow of the Second Coming: American Premillennialism, 1875–1982.* Chicago: University of Chicago Press, 1987.

Williams, Peter. *Popular Religion in America.* Urbana: University of Illinois Press, 1989.

Wuthnow, Robert. "Religious Movements and Counter-Movements in North America." In *New Religious Movements and Rapid Social Change.* James A. Beckford, ed. UNESCO, 1986.

Wuthnow, Robert. *The Restructuring of American Religion: Society and Faith Since World War II.* Princeton, NJ: Princeton University Press, 1988.

Yearbook of the American Churches (New York: National Council of the Churches of Christ in the U.S.A., 1965).

Peale Publications

Peale, Norman Vincent. *The Art of Living.* Nashville, TN: Abingdon Press, 1937.

———. *You Can Win.* Nashville, TN: Abingdon Press, 1938.

———. *A Guide to Confident Living.* New York: Prentice-Hall, 1948.

———. *Not Death at All.* New York: Prentice-Hall, 1948.

———. *The Power of Positive Thinking.* New York: Prentice-Hall, 1952.

———. *The Power of Positive Thinking for Young People.* New York: Prentice-Hall, 1952.

———. *Inspiring Messages for Daily Living.* New York: Prentice-Hall, 1950.

———. *The Coming of the King.* New York: Prentice-Hall, 1956.

———. *Stay Alive All Your Life.* New York: Prentice-Hall, 1957.

———. *He Was a Child.* New York: Prentice-Hall, 1957.

———. *Amazing Results of Positive Thinking.* New York: Prentice-Hall, 1959.

———. *The Tough-Minded Optimist.* New York: Prentice-Hall, 1961.
———. *Adventures in the Holy Land.* New York: Prentice-Hall, 1963.
———. *Sin, Sex and Self-Control.* New York: Doubleday & Company, 1965.
———. *The Healing of Sorrow.* New York: Doubleday & Company and I.B.S., 1966.
———. *Jesus of Nazareth.* New York: Prentice-Hall, 1966.
———. *Enthusiasm Makes the Difference.* New York: Prentice-Hall, 1967.
———. *Treasury of Courage and Confidence.* New York: Doubleday & Company, 1970.
———. *The New Art of Living* (formerly *The Art of Living*). Hawthorn Books, 1971.
———. *Bible Stories.* New York: Franklin Watts, Inc., 1973.
———. *You Can If You Think You Can.* Englewood Cliffs, NJ: Prentice-Hall, 1974.
———. *Favorite Stories of Positive Faith.* Pawling, NY: Foundation for Christian Living, 1974.
———. *Positive Thinking for a Time Like This* (formerly *The Tough-Minded Optimist*). New York: Prentice-Hall, 1961.
———. *The Positive Principle Today.* Englewood Cliffs, NJ: Prentice-Hall, 1976.
———. *The Positive Power of Jesus Christ.* Wheaton, IL: Tyndale House, 1980.
———. *You Can Have God's Help with Daily Problems* (compilation of FCL booklets). Pawling, NY: Foundation for Christian Living, 1980.
———. *Treasury of Joy and Enthusiasm.* Old Tappan, NJ: Fleming Revell, 1981.
———. *Positive Imaging.* Old Tappan, NJ: Fleming Revell, 1982.
———. *The True Joy of Positive Living.* New York: William Morrow, 1984.
———. *Have a Great Day.* Old Tappan, NJ: Fleming Revell, 1984.
———. *Why Some Positive Thinkers Get Powerful Results.* Nashville, TN: Oliver-Nelson Books, 1986.
———. *Power of the Plus Factor.* Old Tappan, NJ: Fleming Revell, 1987.
———. and Ken Blanchard. *The Power of Ethical Management.* New York: William Morrow, 1988.
———. and William Buckley. *The American Character.* Old Tappan, NJ: Fleming Revell, 1988.
———. *How to Be Your Best: A Treasury of Practical Ideas.* Pawling, NY: Foundation for Christian Living, 1990.
———. *Six Attitudes for Winners* (compilation of FCL bookets). Wheaton, IL: Tyndale House, 1989.
———. *Norman Vincent Peale's Favorite Quotations.* New York: Harper & Row, 1990.
———. *This Incredible Century.* Wheaton, IL: Tyndale House, 1991.

Index

RELIGION IN AMERICA SERIES
Harry S. Stout
General Editor

A PERFECT BABEL OF CONFUSION
Dutch Religion and English Culture in the Middle Colonies
Randall Balmer

THE PRESBYTERIAN CONTROVERSY
Fundamentalist, Modernists, and Moderates
Bradley J. Longfield

MORMONS AND THE BIBLE
The Place of the Latter-day Saints in American Religion
Philip L. Barlow

THE RUDE HAND OF INNOVATION
Religion and Social Order in Albany, New York 1652–1836
David G. Hackett

SEASONS OF GRACE
Colonial New England's Revival Tradition in Its British Context
Michael J. Crawford

THE MUSLIMS OF AMERICA
edited by Yvonne Yazbeck Haddad

THE PRISM OF PIETY
Catholick Congregational Clergy at the Beginning of the Enlightenment
John Corrigan

FEMALE PIETY IN PURITAN NEW ENGLAND
The Emergence of Religious Humanism
Amanda Porterfield

THE SECULARIZATION OF THE ACADEMY
edited by George M. Marsden and Bradley J. Longfield

EPISCOPAL WOMEN
Gender, Spirituality, and Commitment in an American Mainline Denomination
edited by Catherine Prelinger

SUBMITTING TO FREEDOM
The Religious Vision of William James
Bennett Ramsey

OLD SHIP OF ZION
The Afro-Baptist Ritual in the African Diaspora
Walter F. Pitts

GOD'S SALESMAN
Norman Vincent Peale and the Power of Positive Thinking
Carol V. R. George